Anonymus

The Victorian Naturalist

Volume 126 (3)

Anonymus

The Victorian Naturalist
Volume 126 (3)

ISBN/EAN: 9783742829894

Manufactured in Europe, USA, Canada, Australia, Japa

Cover: Foto ©berggeist007 / pixelio.de

Manufactured and distributed by brebook publishing software
(www.brebook.com)

Anonymus

The Victorian Naturalist

The
Victorian
Naturalist

126 (3)   June 2009

Biodiversity Symposium Special Issue

The
Victorian
Naturalist

Volume 126 (3) 2009

June

Editors: Anne Morton, Gary Presland, Maria Gibson
Editorial Assistant: Virgil Hubregtse

Biodiversity Symposium Special Issue

ISSN 0042-5184

From the Editors

Birds are an ever-present part of our urban environments. Whether they are ravens or Common Mynahs, in our back yards in seemingly increasing numbers, or powerful owls in bushland parks and reserves, birds are a common sight in the urban expanse. This is perhaps not surprising, since Melbourne is blessed with a wide diversity of natural settings that provide avian wildlife with a range of habitats. Although the extent of wetlands around the metropolitan area have greatly diminished since European settlement, there are still areas of sufficient significance to attract large numbers of water birds from great distances. In addition, throughout the metropolitan area of greater Melbourne there are numerous reserves providing woodland and forest homes for birds. The result is the ongoing presence of species for pleasure and study.

Taking advantage of this, the 2008 Biodiversity Symposium, held by the Field Naturalist Club of Victoria in September, was organised around the theme of 'Birds and the Urban environment'. The papers presented on that occasion covered a wide range of issues related to birds, from the historical record of bird watching and egg collecting around the region, to minimising the impact on bird species of human recreational use of wetlands, to the results of long-term studies of bird numbers in specific environments.

These were interesting papers, deserving of a wider audience than they reached on the day. For that reason alone, the Editors have much pleasure in presenting these papers in this issue of *The Victorian Naturalist*.

Front cover: Little Egret. Photo by Michael Weston

Once in the suburbs: some historical notes on bird watching and collecting in Melbourne

Gary Presland

History and Philosophy of Science
School of Philosophy, Anthropology and Social Inquiry
The University of Melbourne, Victoria 3010

Abstract

The historical records of observations of birdlife within the Melbourne suburban area date from the beginning of European settlement, although they provide little hard evidence regarding local species. The activities of members of FNCV and other individuals, particularly from the early 1880s contributed to a better and more complete picture of local birds. Today, many of the species that were the subject of interest to collectors are rare or endangered. Thanks to the activities of those collectors, however, we can at least refer to a range of records to gain an understanding of the characteristics of the birds that once lived in the suburbs. (*The Victorian Naturalist* 126 (3), 2009, 64-69)

Keywords: Historical records; *The Victorian Naturalist*; suburban bird species

Introduction

The title of this paper is taken in part from an article by a former editor of this journal, GM Ward. In 1966 Ward tried to indicate what he felt could be reconstructed of the pre-European fauna of landscapes subsequently covered by Melbourne's suburbs (Ward 1966). It was an interesting idea at the time, and it is still an interesting idea (Presland 2008). Of course a lot of change has occurred in Melbourne's natural environments in the past 40 years, so it may have been a little easier to do in Ward's day; but even now it remains a fascinating historical objective. Ward's article includes a map of the Melbourne area, showing the distribution of some suburban fauna (Fig. 1). The bulk of the fauna marked on this map are avifauna, so reconsideration of it presents a useful point of introduction to the subject of 'birds in the suburbs'.

One interesting feature of the distribution of suburban fauna displayed in this map is the way in which it is grouped. Being an historical survey, Ward's drawing of the distribution patterns of birds in the suburbs was based on fieldwork previously carried out, often by members of the Field Naturalists Club (FNCV), and published in *The Victorian Naturalist*. The bird species plotted on the map for the area between Doncaster and Ashburton/Caulfield, for example, were those reported by Robert Hall between 1894 and 1898 in the 'Box Hill district', (for more detail see below). Other groupings represent the record of numerous rambles by members of the FNCV in

the Sandringham heathland (see, for example, French 1890; Hart 1890); observations in the old Surveyors' Paddock (Campbell 1899); a report of birds in Royal Park (Anon 1891); and an excursion to the Plenty River (Le Souef 1891).

The work of the early members of the FNCV was important, of course, in developing our knowledge and understanding of birdlife in what are now suburban areas. But observations of birdlife in the Melbourne area were recorded long before the FNCV was formed; indeed, this occurred from the earliest days of European occupation of the area, about 50 years before the formation of the Club in 1880. This brief paper looks back to some of the early observations on local birdlife, the object being to provide brief glimpses of early bird watching and collecting activities in the Melbourne area. These observations lead into the early bird-watching activities of the Field Naturalists Club, reports of some of which are included here. The period thus covered is about 65 years: from the beginnings of European settlement in the Port Phillip region to the end of the 19th century. The intention and hope is that this paper will provide something of an introduction to the subject of a symposium on the theme of urban birds.

19th century perspectives

Before referring to particular reports of birdlife in this area, however, and by way of providing a context to early observations, it is worth mentioning briefly an opinion regarding Australian

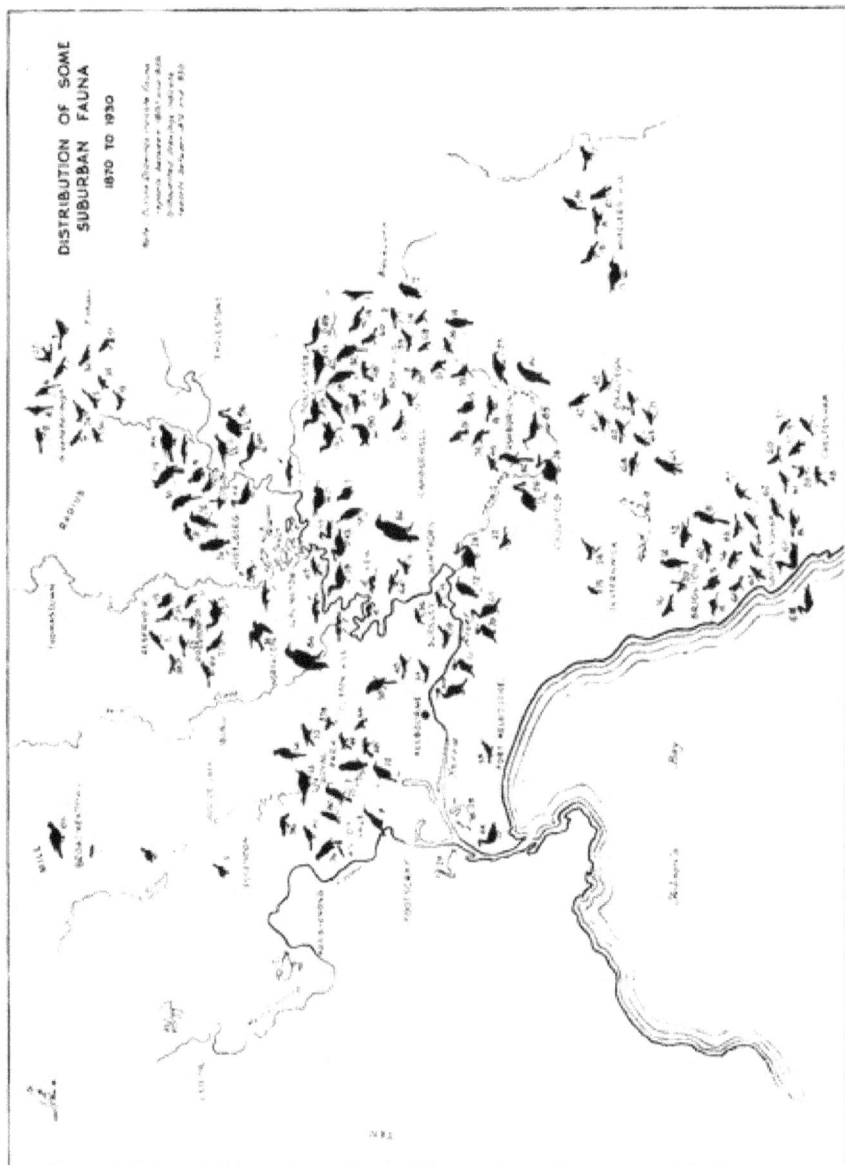

Fig.1. A map showing the distribution of fauna (mostly avian) within the Melbourne metropolitan area, from a paper published in *The Victorian Naturalist* in 1966. The distribution patterns were based primarily on fieldwork done by FNCV members. The original paper included a numbered list of species that corresponded to the numbers on this map.

bird life that was prevalent in the 19th century. This view, expressed in print on a number of occasions (e.g. Martin 1835, 1839), was that the birds of Australia were without song. This is worth mentioning because it was a notion that informed the thinking not only of naturalists but of a wide range of people. Edward Stone Parker, for example, a London teacher who was appointed as an Assistant Protector of Aborigines in 1838, was apparently familiar with Martin's writings on the natural history of Australia. Indeed, he commented on such to his superior GA Robinson in January 1840 while travelling in central Victoria (Presland 1977).

John Cotton, a man of some ornithological experience, was clearly influenced by the same views; in describing the birdlife in the new township of Melbourne, Cotton wrote (1843:20):

> Numerous birds of elegant plumage attract the eye if they do not engage the ear by their musical notes, but they are not all devoid of mellifluous song. Several species of parrakeets [sic] are seen amongst the gum trees, licking their food from the calyx of the blossom ...

The view expressed by Martin probably had two sources: in the first place it was considered generally by 19th century European ornithologists that avian species had begun in the northern hemisphere before radiating out to distant southern locations (Boles 1995, 1997; Barnett 1997). From this perspective it was easy, then, to see Australian birds as inferior, for example lacking in song. It was an expression of the way in which Europeans viewed the Australian fauna as a whole—it was completely foreign and (therefore) strange. The idea that Australian birds did not sing also reflected the opinion of many early settlers, whose ears were accustomed to European birdsong and who thus did not appreciate the very different songs of Australian birds. Actually, it is ironic that people should have thought this, because recent fossil finds suggest that songbirds actually evolved on the Australian continent (Boles 1995, 1997)

Cotton was also an artist and prepared a number of watercolours of birds of the Port Phillip region, drawings that he hoped might serve as illustrations in a book. The book didn't appear in his lifetime, but the paintings survive in a private family collection (McEvey 1974) (Fig. 2). Whilst they do not provide any detailed information regarding the taxonomy or distribution of the illustrated species, for what they are worth these images represent perhaps the earliest pictures of these birds.

Early bird sightings

One of the earliest recorded word pictures referring to birdlife in the Port Phillip area was made within two years of European settlement. Following a short visit to Port Phillip early in 1837, Thomas Winter wrote to the botanist William Swainson reporting *inter alia* on the fauna of the area, including the birdlife (Winter 1898):

> Owls are numerous, and there is a great variety; so are eagles and hawks. White parrots abound. Indeed there is a great variety of this tribe, some very beautiful. Quails are very plentiful, one species being very rare; their colour nearly black, with red spots.

Of course, such fleeting, once-off and non-specific observations are of little scientific value. But they do serve as a reminder of a couple of considerations regarding historical observations: season and circumstance. The earliest European visit to the Melbourne area was in February 1803, by a party intent on making an assessment of the area for the purposes of permanent settlement. Birdlife was of less consideration than good soil, timber and fresh water. Occasional references to birds are made, but there is nothing of any import regarding avian species, in the written account of the expedition (Flemming 1879).

However, once Europeans permanently occupied the area, observations on the local birdlife began to be recorded regularly. Birds were noted in all of the various environments around the Port Phillip area; initially, of course, most observations were made in the vicinity of the town. For example, West Melbourne Swamp in the period before the 1860s, when reclamation of the wetland area was begun, was described (Mattingley 1916:83) as one where:

> graceful swans, pelicans, geese, black, brown and grey ducks, teal, cormorants, water hen, sea gulls ... disported themselves; while curlews, spur-winged plover, cranes, snipe, sandpipers and dotterals either waded in the shallows or ran along its margin; and quail and stone plover ... were very plentiful ...

Most of these birds were to be found across the entire area of the Yarra estuary for some part of the year. Along the Yarra itself, birdlife was abundant. In the early days of the settle-

Fig. 2. A Ground Thrush (now Bassian Thrush) *Zoothera lunulata* as drawn by John Cotton, in 1843.

ment, travellers on the river noted particularly the Nankeen Night-Heron (sometimes called Rufous Night-Heron) *Nycticorax caledonicus* in the adjacent shrubs and bushes (Backhouse 1843). The wetlands in low-lying areas in Footscray, immediately to the west of the Maribyrnong River, provided rich feeding grounds for wild geese, as well as places to breed. The birds flocked in such numbers that on occasions the sky was darkened by their flight and the area became known as Gosling Flat (Anon 1960).

Emus were once a common sight on the open grassy plains around Port Phillip. When was the last time anybody was approached by an Emu in Dandenong? No doubt it hasn't happened recently, but in the 1840s the species was so common in the locality that settlers had trouble keeping emus out of their tents. In a letter to GA Robinson, William Thomas (1839) observed that the alluvial flats around the Dandenong area

... were the resort of the emu and kangaroo, so numerous were these animals but 18 months back that it was not uncommon for emus to come up to the hut and tent door but not so now.

Almost exactly a year later he wrote that 'there is but little game within 20 miles of Melbourne'

(Thomas 1840), and by the mid 1850s Horace Wheelwright could comment that 'an emu ... killed within forty miles of the town would be a matter of history'. Wheelwright was a professional shooter, who made his living by providing game for the tables of the growing town, so to some extent he was responsible for the fact that game birds had become scarce. In his memoir of 1861 (1979) Wheelwright listed more than 180 species he had sighted within a 40 mile (64 km) radius of Melbourne. Wheelwright 'bagged' many of these species but unfortunately his memoirs do not provide precise details regarding the locations where the birds were shot.

On the plains to the north and west of Melbourne, Australian Bush turkeys also were plentiful in the early days of European settlement (Edwards 1854). It was not uncommon to see as many as 30 birds in a single flock, despite the belief by some bushmen that these birds were shy in their habits (Wheelwright 1879).

FNCV activities

No doubt the influx of large numbers of settlers to the region, particularly following the discovery of gold in the colony in 1851, had detrimental effects on local birdlife. However, even as late as 1890, there were at least 13 avian species still living and breeding in the area of Royal Park (Anon 1891).

Further out from the city there were better prospects for observing birds. In the Box Hill district, for example, there were numerous excursions by members of the FNCV to document and collect birds and their eggs, particularly following the opening of a rail line to Lilydale in 1883. As a result of these fieldtrips, between 1896 and 1900 there were nine papers published in *The Victorian Naturalist*, all by Robert Hall (Hall 1894, 1896a, 1896b, 1896bc, 1897a, 1897b, 1898a, 1898b, 1898c, 1898d). The author refers in the titles of these pieces to the 'Box Hill District' but in fact the articles focused on a greater area than that: the occurrence of birds was recorded in a roughly triangular area that extended from Doncaster in the north to Oakleigh in the south, and as far to the east as Bayswater (Hall 1894). Of particular interest were the valleys of Koonung Koonung, Blackburn and Gardiners Creeks. The wooded nature of that area ensured that particular species were likely to be represented, including varieties of cuckoos, honeyeaters, owls, swallows and pardalotes. Over the series of articles, Hall details all of the birds that comprised the major families; today this is an invaluable record. Summarising the data collected over a period of five years, Hall noted that of the 106 native species of birds observed in the area, 36 resided there all year, and the other 70 were migrants, generally arriving in spring to nest and rear their young.

Other members of the FNCV were active at the same time, in these and other areas. AJ Campbell, who was first elected to the Club in 1881 and was a Committee Member in the latter part of the 1880s, maintained his bird observation and collecting activities over many decades and across the entire continent, including Melbourne and its suburban area. He was an active participant at Club meetings, an exhibitor of eggs and bird skins and a regular contributor to *The Victorian Naturalist* until his death in 1926. To give him his due, Campbell was also an active proponent, through the FNCV, of the protection of native species (Anon 1885). His magnum opus *Nests and eggs of Australian birds*

was published in two volumes in 1900 and contains interesting observations and remarks on native species. Some of these remarks are worth relating here in the context of avifauna in the Melbourne suburban area.

On what was originally called the Warty-faced Honeyeater (now Regent Honeyeater) *Anthocaera phrygia*, Campbell wrote (1900:382):

> I recollect one season in November—1868 or 1869—when these birds were plentiful in the neighbourhood of Oakleigh and Murrumbeena, where we secured as many of their beautifully-constructed bark-made nests, and lovely rich salmon-coloured eggs as we needed.
>
> I myself witnessed ... once at Doncaster, 2nd November 1886 ... a flock of about fifty swept past me across a valley.

On the Helmeted Honeyeater: in 1884 Campbell was informed by fellow FNCV member AW Milligan that he had seen a large flock of the species in the vicinity of Olinda Creek near Lilydale. As it happened, the very first camp-out by the FNCV, organised in part by Campbell (Barnard 1906), was to take place in that area on the long weekend in November to mark the birthday of the Prince of Wales. During the weekend a nest containing an egg of the Helmeted Honeyeater was discovered by the bird party, led by Campbell. He later related the chain of events that ensued (Campbell 1900:400):

> ... the honour fell to the late Mr. W Hatton of detecting the first nest, with the rare honeyeater sitting. The nest was situated at a height of about twenty feet, and was suspended to an outstretched branch of a hazel overhanging the creek. With what ecstasy of delight the small tree was ascended! The handsome bird still retained possession of its nest. With Mr. Hatton's assistance, I all but had my hands on the coveted prize, when, without a moment's warning, crash went the tree by the root, and all – the two naturalists, tree, bird, nest and eggs – went headlong into the stream beneath.

The account goes on to relate how the nest and eggs were retrieved intact, because the bird continued to sit on it, despite being dunked in the creek. Three other nests, with eggs, were found subsequently in the area, and all the eggs probably ended up in either the National Museum of Victoria or in the cabinet of a private collector.

Conclusion

These were the customary practices and activities of members of the FNCV, as well as those of a lot of other bird watchers and collectors of the day. It would be wrong, certainly, to suggest that such activities were the sole, or indeed the major cause of the fact that this bird (Victoria's faunal emblem), as well as the Regent Honeyeater, is listed now as threatened (DSE 2008). Doubtless other factors, for example habitat destruction, were at play, contributing significantly to threaten and endanger the survival of these two and many other bird species. While we might take a somewhat different approach in our studies of such species today, this is perhaps due in part to the fact that there aren't as many of these birds around any more. In the past people obviously found them attractive, as we do today. But it seems that being attractive was their undoing, and the poor creatures were loved almost to extinction!

Although avian species were commented on from the earliest days of European settlement in the Melbourne area, the records are too lacking in essential detail to be of great value. As with any other animals, what is needed is closer, repeated observation of members of the species in the field, particularly over protracted periods. The beginnings of such measured study came with the formation and subsequent field activities of the FNCV. It is through the records created by these observers, published in the pages of *The Victorian Naturalist* and elsewhere, that we can begin to understand the characteristics of the birds that once lived in the suburbs.

References

Anon. (1885) Report of Club activities *The Victorian Naturalist* 2, 13.

Anon. (1891) Native birds breeding in the Royal Park in 1890 *The Victorian Naturalist* 7, 155.

Anon. (1960) *Footscray's first 100 years: the story of a great Australian city.* (Footscray: *The Advertiser*/ Footscray City Council)

Backhouse J (1843) *Narrative of a visit to the Australian colonies.* (London: Hamilton, Adams & Co)

Barnard FGA (1906) The first quarter of a century of the Field Naturalists Club of Victoria *The Victorian Naturalist* 23, 63-77.

Barnett A (1997) *New Scientist*, 28 June 1997.

Boles W (1995) The world's oldest songbird *Nature* 374, 21-22.

Boles W (1997) Fossil songbirds (Passeriformes) from the Early Eocene of Australia *Emu* 97, 43-47.

Campbell A (1899) List of birds observed at Burnley *The Victorian Naturalist* 16, 49-55.

Campbell AJ (1900) *Nests and eggs of Australian birds* (Melbourne: The Author)

Cotton J (1843) *The Correspondence of John Cotton: Victorian pioneer 1842-1849.* 2 ed. Ed. C Mackeness. Australian Historical Monographs Vol. 28 (Dubbo: Review Publications. 1978)

DSE (2008) *Flora and Fauna Guarantee Act 1988*, Threatened List 2008

Edwards H (1854) Letter In Estate and family papers of Sir Anselm Guise. Gloucester Records Office

Flemming J (1879) A journal of the explorations of Charles Grimes, Acting Surveyor-General of New South Wales. In *Historical Records of Port Phillip*, pp. 12-39. Ed JJ Shillinglaw. (Melbourne: Victorian Government)

French C (1890) A ramble through the heath-ground from Oakleigh to Sandringham *The Victorian Naturalist* 7, 71-75.

Hall R (1894) Birds of Box Hill District *The Victorian Naturalist* 11, 91-92.

Hall R (1896a) Box Hill birds in July 1896 *The Victorian Naturalist* 13, 103-107.

Hall R (1896b) Notes on the bird fauna of the Box Hill district. Part 1 *The Victorian Naturalist* 12, 127-134.

Hall R (1896c) Notes on the bird fauna of the Box Hill district. Part 2 *The Victorian Naturalist* 12, 143-146.

Hall R (1897a) Notes on the bird fauna of the Box Hill district. Part 3 *The Victorian Naturalist* 14, 53-58.

Hall R (1897b) Notes on the bird fauna of the Box Hill district. Part 4 *The Victorian Naturalist* 14, 69-73.

Hall R (1898a) Notes on the birds of Box Hill district. Part 1 *The Victorian Naturalist* 15, 70-72.

Hall R (1898b) Notes on the birds of Box Hill district. Part 2 *The Victorian Naturalist* 15, 75-80.

Hall R (1898c) Notes on the birds of Box Hill district. Part 3 *The Victorian Naturalist* 15, 127-130.

Hall R (1898d) Notes on the birds of Box Hill district. Part 4 *The Victorian Naturalist* 15, 156-159.

Hart JS (1890) Excursion to Cheltenham *The Victorian Naturalist* 7 (3): 85-86.

Le Souef, D (1891) Report of a trip to Plenty River, 20 December 1890. *The Victorian Naturalist* 8, 7-8.

McEvey A (1974) *John Cotton's birds of the Port Phillip district of New South Wales, 1843-1849* (Sydney: Collins)

Martin RM (1835) *History of the British colonies*, (volume IV) (London: James Cochrane & Co)

Martin RM (1839) *History of Austral-Asia: comprising New South Wales, Van Diemen's Island, Swan River, South Australia, &c.* 2 ed (London: Whittaker)

Mattingley A (1916) The early history of North Melbourne, Part 1. *Victorian Historical Magazine* 10, 80-92.

Presland G (Ed) (1977) Journals of George Augustus Robinson, January to March 1840. *Records of the Victorian Archaeological Survey* 5

Presland G (2008) *The place for a village: how nature has shaped the city of Melbourne* (Melbourne: Museum Victoria)

Thomas W (1839) Letter to G.A. Robinson, 6 August 1839 In Papers of William Thomas. Mitchell Library Sydney.

Thomas W (1840) Letter to G.A. Robinson, 26 August 1840 In Papers of William Thomas. Mitchell Library Sydney, .

Ward GM (1966) Once in the suburbs. *The Victorian Naturalist* 83, 157-167

Wheelwright HW (1979) *Bush wanderings of a naturalist* (Facs ed) (Melbourne: Oxford University Press)

Winter T (1898) Notes on Port Phillip In *Letters from Victorian pioneers*, pp. 275-279 Ed TF Bride (Melbourne: Trustees of the Public Library)

Received 20 November 2008; accepted 23 April 2009

Food resources and urban colonisation by lorikeets and parrots

Alan Lill

Wildlife Ecology Research Group
School of Biological Sciences, Monash University, Clayton Campus, Victoria 3800

Abstract

Several native bird species have recently successfully colonized many Australian cities. The presence of some of them may be largely beneficial, but their urban ecology is poorly understood. We conducted short-term studies of the foraging ecology of Rainbow and Musk Lorikeets and Red-rumped Parrots in Melbourne parklands to help fill this knowledge gap. The nectar (and/or pollen) of six eucalypt species, mostly not native to the Melbourne area, strongly dominated the lorikeets' diet year-round. The key eucalypt species variously flowered for 80-100% of winter and 72-84% of summer. In winter, 80% of the Red-rumped Parrot's diet comprised the abundant seeds of four exotic grasses and herbs. There was little evidence of significant inter-specific competition, particularly through aggressive interference, for any of the lorikeets' or parrots' urban food resources. Thus a critical factor facilitating urban colonization by these birds seems to be that, collectively, ornamental eucalypts planted last century, turf grasses commonly occurring on sports grounds and in parks and common weeds provide abundant food resources in Melbourne's parklands that are broadly similar to those of their non-urban habitats. Moreover, exploitation of these resources by other urban birds seems to be fairly limited. (*The Victorian Naturalist* 126 (3), 2009, 70-72)

Keywords: Lorikeets, parrots, urban colonization, diet, eucalypt nectar, grass and herb seeds, inter-specific aggression

Several native bird species that appear to be increasing in abundance and expanding their geographic ranges have recently colonised many of Australia's major cities. Noisy Miners *Manorina melanocephala* and Pied Currawongs *Strepera graculina* are suspected of adversely affecting whole suites of native bird species in some of our cities (Low 2002), but for other so-called 'urban adapters' (Blair 2001), negative impacts on cohabiting native birds are less obvious and perhaps even non-existent. The reality is that the urban ecology of many of these native, invasive birds is poorly known. We need to bridge this knowledge gap in order to understand what causes and facilitates these urban invasions and to properly evaluate their consequences for urban biodiversity conservation. My research group has been addressing this task by conducting single-season, 'snapshot' studies of the foraging ecology of several of these native, urban invasive species, including that of two lorikeets and a parrot, in Melbourne parkland.

The Rainbow Lorikeet *Trichoglossus haematodus* re-established itself in Melbourne in the 1970s after a prolonged absence, or perhaps a period of extremely low abundance, since the late 1800s (Crome and Shields 1992). It is now abundant and widespread in the city all year round (Shukuroglou and McCarthy 2006).

Musk Lorikeets *Glossopsitta concinna* have also increased in abundance in Melbourne since the 1970s (Higgins 1999) and recently their presence has become less seasonal. The range of the Red-rumped Parrot *Psephotus haematonotus* has expanded into south-eastern Australian coastal cities in the last 60 years (Higgins 1999) and for some time now it has been common in some Melbourne parks, particularly in winter. Our studies of these three species have documented their diet and the seasonal availability of their main food resources and examined whether other bird species appear to be significant interference competitors for their food resources in the city (Lowry and Lill 2007; Smith and Lill 2008; Stanford and Lill 2008). The present account draws on these investigations to address the issue of how food availability might have influenced the colonising of Melbourne by these species.

Almost all (99%) foraging by Rainbow and Musk Lorikeets was conducted in the tree canopy stratum. All winter foraging was performed whilst perching upright (59-60% of observations) or hanging upside-down; summer foraging behaviour was similar, although hanging upside-down was a little less common. Eucalypt nectar (and/or pollen) strongly dominated the lorikeets' diet year-round, being the item consumed in 86-97% of over 6 000 foraging observations; seeds, fruit and invertebrates were minor dietary components. Most of the nectar/pollen was obtained from six eucalypt species. The two most prominent species were Spotted Gum *Corymbia maculata*, which accounted for 27-29% of nectar/pollen foraging in winter and 12% in summer, and Red Ironbark *Eucalyptus sideroxylon*, which accounted for 26-27% of nectar/pollen foraging in winter and 17-22% in summer. Yellow gum *Eucalyptus leucoxylon* was also an important nectar/pollen source for the lorikeets in both seasons and Southern Blue-gum *Eucalyptus globulus* accounted for 7-13% of their nectar/pollen foraging in winter. Most of the eucalypts exploited are not native to the Melbourne area. In summer, 61-62% of the lorikeets' nectar/pollen was obtained from these introduced eucalypts and in winter the percentage was even higher (72-84%). Our phenological studies showed that the key eucalypt food plant species in winter variously flowered for 80-100% of the time and the key species in summer for 67-95% of the time.

Collectively, six other native bird species fed on 10 of the 16 winter food plant species of the lorikeets, particularly on the nectar/pollen of Spotted Gum, Red Ironbark and Yellow Gum. Six other native bird species exploited a total of 13 of the lorikeets' 33 summer food plants too, particularly the nectar of Red Ironbark and Sugar Gum *Eucalyptus cladocalyx*. All these potential competitors were honeyeaters (Meliphagidae), cockatoos (Cacatuidae) or parrots (Psittacidae). However, only Noisy Miners and Red Wattlebirds *Anthochaera carunculata* were significant exploiters of the lorikeets' nectar resources and even this exploitation was only at 17% and 4% of the lorikeets' winter and summer usage rates, respectively. Consistent with this pattern, lorikeets were rarely involved in inter-specific aggression over food (winter 0.6 and summer 3.5 interactions per observation week) and most of the interactions observed had little negative effect on the lorikeets' foraging behaviour. For example, being displaced >2 m but not out of the feeding site was the most common outcome for both lorikeets in summer (41% and 59% of interaction outcomes for Rainbow and Musk Lorikeets, respectively). Noisy Miners were involved in 85% of these summer inter-specific encounters.

Red-rumped Parrots occurred at mean population densities of ~ 0.5-3 per ha and in flocks of 1-139 (mean =10) in Melbourne parks in winter. They fed mainly on the ground, less than 2% of foraging occurring in trees. The diet comprised mainly the seeds (78% of foraging observations) and, to a lesser extent, the buds (11%) of thirteen plant species, mainly exotic grasses and herbs. A few flowers of herb species were also consumed. However, just four exotic plant species collectively provided 83% of the diet. Annual Bluegrass *Poa annua* and Kikuyu Grass *Pennisetum clandestinum* seeds together comprised just over half of the diet and the seeds of two herbs commonly regarded as weeds, Knotweed *Polygonum arenastrum* and Chickweed *Stellaria media*, accounted for a further 24% of food items consumed. Our measurements showed that this seed resource was abundant throughout winter. Intriguingly, it was just as available in sites not occupied by, but superficially suitable for red-rumps, as in sites used by them. Thus the mean proportional availability of Annual Bluegrass, based on estimates of percentage cover, was 28.7% in occupied sites and 26.9% in unoccupied sites. However, the occupied sites may have provided better protection for roosting red-rumps diurnally and nocturnally, because they had more tall trees and dense canopy cover. Again, the negligible amount of aggression observed between red-rumps and cohabiting bird species over food (6 encounters in 40+ hours of observation of foraging birds) had little apparent negative effect on the parrot's foraging behaviour; they were either displaced < 5 m or showed no overt response.

Melbourne's parks apparently usually provide an abundant nectar/pollen supply for Rainbow and Musk Lorikeets. A major reason for this appears to be the planting of over 120 eucalypt species, many of which are not native to the area, as ornamentals last century (Beer *et al.* 2001). This diversity, perhaps augmented by the urban heat sink effect and a high soil moisture content resulting from artificial watering (Neil and Wu 2006), has apparently resulted in longer flowering seasons overall and hence an abundant, year-round nectar supply for urban lorikeets (Fitzsimons *et al.* 2003). There seems to be only limited inter-specific interference competition with other birds for this resource. Exotic grasses, commonly occurring in turf in parks and sports fields, along with common weeds provided an abundant seed resource for red-rumps throughout winter in Melbourne.

Critically, Kikuyu Grass, which provided 22% of the Red-rumped Parrot's winter diet, produces seeds prolifically even when regularly mowed (Huff 2002). Exotic and native ornamental trees provide suitable roosts for red-rumps in the city's parks and, as with the lorikeets, there did not seem to be significant inter-specific interference competition for food resources. Melbourne provides food resources closely approximating those in the three parrots' non-urban habitats (Higgins 1999), so dietary flexibility has not been a pre-requisite for urban colonisation by these birds. They may compete with other native animals for tree-hollow nest sites, but otherwise their urban presence seems to be mostly beneficial.

Our short-term studies need to be replicated in additional years, given the known annual variation in eucalypt flowering phenology (Law *et al.* 2000) and possible effects of drought on seeding grasses. Lorikeets also need to be studied in gardens and streetscapes, which they use extensively. The inter-specific competition issue requires further evaluation through a more comprehensive examination of the entire diet of possible competitor species. Finally, we need to see if the Melbourne picture holds for other cities colonised by these parrots and to identify the factors in the parrots' non-urban environment that have led to the urban niche being exploited by these birds.

Acknowledgements

I thank Hélène Lowry, Justine Smith and Lauren Stanford for doing most of the hard work in these studies. An anonymous referee made some helpful comments on the manuscript.

References

Beer R, Frank S and Waters G (2001) Overview of street tree populations in Melbourne - turn of the 21st century Horticultural Project Report. Burnley College, University of Melbourne

Blair RB (2001) Birds and butterflies along urban gradients in two ecoregions of the U.S. In *Biotic Homogenization*, pp 33-56. Eds JL Lockwood and ML McKinney. (Kluwer-Norwell, MA)

Crome F and Shields J (1992) *Parrots and Pigeons of Australia*. (Angus and Robertson, Sydney)

Fitzsimons JA, Palmer GC, Antos MJ and White JG (2003) Refugees and residents: densities and habitat preferences of lorikeets in urban Melbourne. *Australian Field Ornithology* 20, 2-7

Higgins PJ (ed) (1999) *Handbook of Australian, New Zealand and Antarctic Birds. Vol. 4: Parrot to Dollarbird*. (Oxford University Press: Melbourne)

Huff T (2002) Basic biology of annual bluegrass (*Poa annua* L.) In *Annual Bluegrass (Poa annua L.) – Biology, Management and Control*, pp. 37-42 Ed D Aldous (University of Melbourne: Melbourne)

Law B, Mackowski C, Schoer L and Tweedie T (2000) Flowering phenology of myrtaceous trees and their relation to climatic, environmental and disturbance variables in northern New South Wales. *Austral Ecology* 25, 160-178

Low T (2002) *The New Nature* (Penguin: Melbourne)

Lowry H and Lill A (2007) Ecological factors facilitating city-dwelling in red-rumped parrots. *Wildlife Research* 34, 624-631.

Neil, K. and Wu, J. (2006) Effects of urbanization on plant flowering phenology: a review. *Urban Ecosystems* 9, 243-257.

Shukuroglou P and McCarthy M (2006) Modelling the occurrence of rainbow lorikeets (*Trichoglossus haematodus*) in Melbourne. *Austral Ecology* 31, 240-253.

Smith J and Lill A (2008) Importance of eucalypts in exploitation of urban parks by Rainbow and Musk Lorikeets. *Emu* 108, 187-195.

Stanford L and Lill A (2008) Out on the town: winter feeding ecology of lorikeets in urban parkland. *Corella* 32, 49-57.

Received 6 November 2008; accepted 5 February 2009

Red-rumped parrot *Psephotus haematonotus*. Photo by Virgil Hubregtse.

Surviving urbanisation: maintaining bird species diversity in urban Melbourne

John G White[1], James A Fitzsimons[1,2], Grant C Palmer[1,3], and Mark J Antos[1,4]

[1] School of Life and Environmental Sciences, Deakin University.
221 Burwood Highway, Burwood, Victoria 3125. Email: john.white@deakin.edu.au
[2] The Nature Conservancy, 60 Leicester Street, Carlton, Victoria 3053 Email: jfitzsimons@tnc.org
[3] Centre for Environmental Management, School of Science and Engineering,
University of Ballarat, P.O. Box 663, Ballarat, Victoria 3353. Email: g.palmer@ballarat.edu.au
[4] Research Branch, Parks Victoria, Level 10, 535 Bourke Street, Melbourne, Victoria 3000.
Email: mantos@parks.vic.gov.au

Abstract

The relationships between vegetation and bird communities within an urban landscape are synthetised, based on a series of studies we conducted. Our studies indicate that streetscape vegetation plays an important role in influencing urban bird communities, with streetscapes dominated by native plants supporting communities with high native species richness and abundance, while exotic and newly-developed streetscapes support more introduced bird species and fewer native bird species. Native streetscapes can also provide important resources for certain groups of birds, such as nectarivores. Our research has also revealed that urban remnants are likely to support more native bird species if they are larger and if they contain components of riparian vegetation. Vegetation structure and quality does not appear to be as important a driver as remnant size in determining the richness of native bird communities. Introduced birds were shown to occur in remnants at low densities, irrespective of remnant size, when compared to densities found in streetscapes dominated by exotic vegetation. We discuss our results in terms of practical planning and management options to increase and maintain urban avian diversity and conclude by offering suggestions for future fields of research in terms of urban bird communities. (*The Victorian Naturalist* 126 (3), 2009, 73-78)

Keywords: urbanisation; bird assemblages; remnant vegetation; streetscapes; riparian zones

Introduction

Increasing urbanisation is a major threat to biodiversity, and as such there is considerable interest in mitigating its impacts on natural systems. The process of urbanisation converts natural and/or agricultural environments into 'novel', yet diverse, environments consisting of buildings, roads, streetscapes, open space and remnants of native vegetation. Research in Australia that documents how biodiversity responds to urbanisation is limited (for reviews see Lunney and Burgin 2004, Garden *et al.* 2006), and thus we have limited knowledge on how to manage urban environments to maintain biodiversity. Nonetheless, it is encouraging to see an increase in research interest in urban biodiversity, and particularly urban bird ecology, in recent years. Areas of research in Melbourne range from habitat preferences of bird communities (our work – see below), to human disturbance impacts (e.g. Platt and Lill 2006; Price and Lill 2008; Weston *et al.* 2009) to single species studies (e.g. Lowry and Lill 2007). This has been complemented by work in other Australian cities (e.g. Parsons *et al.* 2003, 2005; Daniels and Kirkpatrick 2006; Young *et*

al. 2007) and a burgeoning international literature, as well as the appearance of specialised journals (e.g. *Urban Ecosystems, Landscape and Urban Planning*).

This paper utilises our previous research investigating the impacts of urbanisation on bird communities (i.e. Fitzsimons *et al.* 2003; White *et al.* 2005; Antos *et al.* 2006; Palmer *et al.* 2008) to highlight key findings and implications for conserving and promoting diversity in urban bird assemblages. Our research has examined the influence of streetscape vegetation on bird assemblages, the distribution of introduced birds within urban remnants and the key drivers of native avian species richness and composition within remnants. In this paper, we provide a synthesis of our findings and management recommendations.

Summary of methods and results

All research described in this paper was conducted in the eastern and south-eastern suburbs of Melbourne, within a 30 km radius of the CBD, during 2002-2004.

Birds in streetscapes

To conduct this study, the urban areas of Melbourne were broadly divided based on the dominant streetscape trees, and then compared to patches of remnant vegetation (also in the urban environment). The three streetscape types were those dominated by established native trees (not necessarily indigenous), streetscapes dominated by established exotic trees, and streetscapes in new suburbs where there was limited vegetation. In each of the four site types there were nine replicate sites, yielding a total of 36 sites. One hectare transects were established at each site and surveyed on three separate occasions. Each bird species was recorded and the average number of individuals of each species was determined in order to provide a measure of relative abundance. For a detailed description of the study, see White *et al.* (2005).

In this study we recorded 60 native species and seven introduced species. The bird community composition differed between each of the different types of sites. The richness of native bird species differed considerably between site types, with the lowest richness occurring in streetscapes with exotic trees and in new suburbs (Fig. 1). Both remnants and established native streetscapes had high richness of native species. A similar trend was observed for the abundance of native birds, with remnants and native streetscapes having higher abundances than exotic streetscapes and new developments (Fig. 1). The richness of introduced bird species was associated with the type of site, with remnants having low richness compared to all the streetscape types. The major difference, however, was observed when investigating the abundance of introduced birds. The abundance of introduced species was lowest in remnants, increased in native streetscapes, and was highest in exotic streetscapes. New streetscapes had intermediate levels between the exotic and native streetscapes, but were not significantly different from either (Fig. 1).

Another way of investigating community complexity is to compare the number of different feeding guilds represented in different types of sites. Overall, the highest numbers of guilds were represented in the remnant vegetation. Native streetscape areas were also well represented and supported most guilds found in remnants. There was, however, a consider-able drop in the number of feeding guilds, and thus a drop in community complexity, in exotic streetscapes and new developments (Fig. 1). The major difference in guild composition between the native streetscapes and the exotic and new streetscapes was the reduction in insectivores and nectarivores in exotic and new streetscapes.

Some native bird species (e.g. lorikeets) were recorded in very high abundances in native streetscapes and appear to have been favoured by the planting of native, but non-indigenous, eucalypts (Fitzsimons *et al.* 2003). These non-indigenous eucalypts are generally more profuse flowerers than indigenous eucalypts, and lorikeets have been shown to preferentially select them in urban areas (e.g. Smith and Lill 2008, Stanford and Lill 2008).

Overall, these findings, and similar recent findings in Adelaide by Young *et al.* (2007), suggest that the type of streetscape planting has a considerable influence on bird communities. Streetscapes supporting native vegetation, be it remnant or planted, support richer bird assemblages dominated by native species, and provide effective 'nature strips' for at least some native bird species.

Birds in remnant vegetation

Thirty-nine remnants of native vegetation were surveyed for birds in this study. The remnants ranged in size from 1 ha to 107 ha. These sites were surveyed four times each during both the breeding season and non-breeding season for both native and introduced bird species. We excluded aquatic bird species from any comparisons because many remnants did not have aquatic habitats (for detailed methods see Antos *et al.* (2005) and Palmer *et al.* (2008)). Overall, introduced birds did not demonstrate any major trends in abundance and distribution in urban remnant vegetation. Whilst the composition changed with increasing remnant size, the relative abundance of introduced birds was largely unaffected by remnant size (Antos *et al.* 2005). In general the abundance of introduced birds was very low in urban remnants when compared to streetscape vegetation.

In this study 79 native woodland bird species were recorded (see Palmer *et al.* 2008 for details). The richness of birds in remnants was strongly influenced by the size of the remnant patch (Fig. 2). In general, almost all remnants had a base bird community

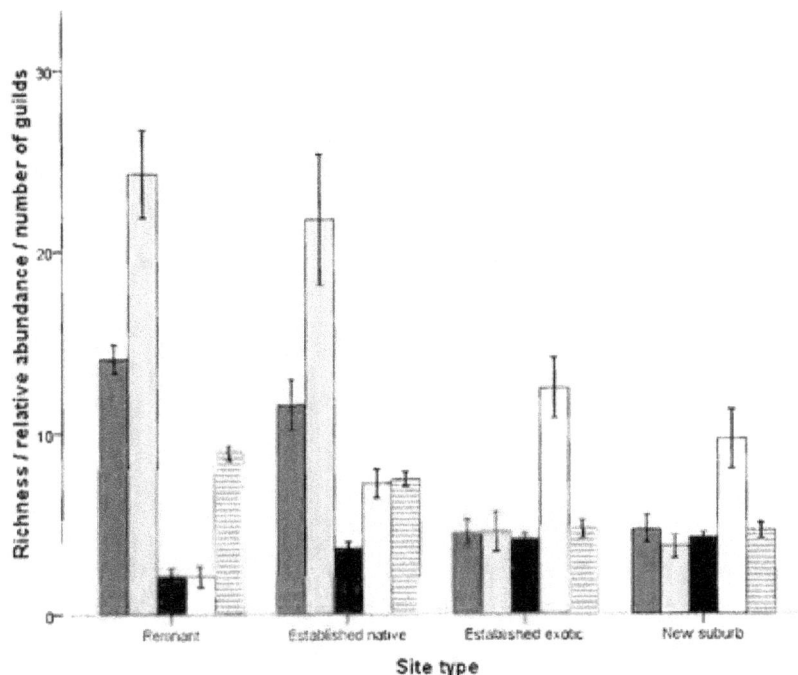

Fig. 1. The influence of different urban sites on bird community composition (Mean ± 1SE). Dark grey bars = number of native species; light grey bars = relative abundance of native species (birds/ha); black bars = number of introduced species; white bars = relative abundance of introduced species (birds/ha); horizontally striped bars = number of feeding guilds. After White *et al.* (2005).

consisting of nine species, these being Red Wattlebird *Anthochaera carunculata*, Rainbow Lorikeet *Trichoglossus haematodus*, Eastern Rosella *Platycercus eximius*, Australian Magpie *Cracticus tibicen*, Spotted Pardalote *Pardalotus punctatus*, Little Raven *Corvus mellori*, Brown Thornbill *Acanthiza pusilla*, Noisy Miner *Manorina melanocephala* and Grey Butcherbird *Cracticus torquatus*. All these species were also well represented in streetscape sites, which suggests they are reasonably tolerant of the urban matrix. All native species recorded within the remnants were classified into categories ('all species', 'urban tolerant', 'urban sensitive', 'ground foragers', 'shrub foragers', 'canopy foragers' and 'migrants') and assessed against a series of parameters associated with the remnant patches (e.g. remnant size, amount of surrounding vegetation, vegetation life-form cover etc). All these different groupings (excluding

'urban tolerant' species) showed strong positive relationships between richness and the size of the remnant, adding further support for the finding that the size of a remnant is critical for bird diversity (Table 1). The richness of most groupings of birds was not significantly affected by the quality of either the ground vegetation or the canopy and shrub layer (Table 1). With the exception of the richness of migrant species, most species were not influenced by the amount of remnant vegetation in a 500 m radius around each remnant. Other than remnant size, the only aspect of the remnant that affected richness of species was the amount of riparian vegetation within the remnant. Riparian vegetation may be more productive for birds, but also may be providing connectivity between remnants, as remnant vegetation often occurs along creeklines in the urban landscape studied.

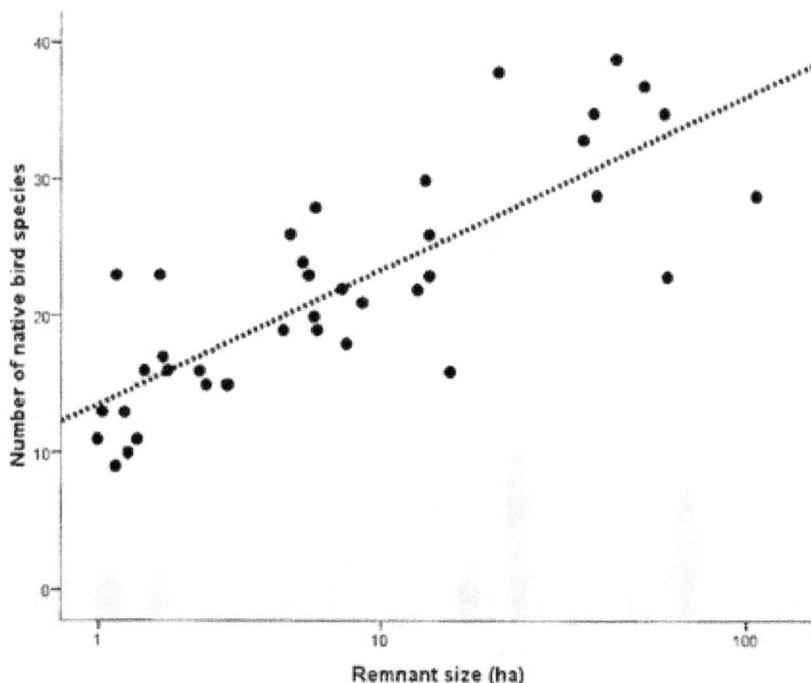

Fig. 2. Relationship between native woodland bird species richness and remnant size based on the 39 remnants examined by Palmer *et al.* (2008).

Table 1. Relative strength of relationship between avian ecological groups and habitat variables within urban vegetation remnants. +++ = strong positive relationship, ++ = moderate positive relationship, + = weak positive relationship. Blank cells indicate no detectable differences. After Palmer *et al.* (2008).

Type of species	Remnant size	Canopy shrub complexity	Ground layer complexity	% vegetation in surrounding landscape	% riparian vegetation
All species	+++				++
Urban tolerant species	+				
Remnant reliant species	+++				+
Ground foragers	+++				
Shrub foragers	++	++			
Canopy foragers	+++				++
Migrants	+++			++	

Practical opportunities for planning and management

The findings outlined above provide directions for both urban planners and residents to increase urban bird diversity in Melbourne and other urban areas. Some key principles include:

- Protect all remnants of native vegetation, which are the base for urban biodiversity;
- Initially focus on increasing the size of remnants by revegetation or reservation of the largest area available (where applicable);
  - Increasing the size of the remnant appears more important than improving vegetation quality;
- Turn streetscapes into 'nature strips';
  - Plant native trees and replace, or at least supplement, exotic trees with native trees;
  - Reduce exotic grass cover and replace with shrubs and native ground cover to enhance nature strips for native birds (see Parsons 2007);
- Increase native vegetation in residential gardens and areas of community open space.

There are a number of opportunities in Melbourne's growth corridors (e.g. Cranbourne-Pakenham growth corridor) to plan viable reserves within existing and proposed residential developments, and we need to make the most of these. Making the findings of urban biodiversity research accessible to key stakeholders and planners will be important for this to happen.

A review of public land use by the Victorian Environmental Assessment Council (VEAC) is currently under way across urban Melbourne (see <www.veac.vic.gov.au> for more details). This body, and its predecessors, the Land Conservation Council and Environment Conservation Council, have been responsible for the creation of most of the present day park and reserve system across the state but these bodies have not previously made recommendations concerning public land use in Melbourne. Many important larger remnants occur in areas not currently reserved and/or managed for conservation, such as on freeway reservations. The results of our research suggest that consolidating larger areas of native vegetation in single and, if possible, connected systems, will provide for a greater diversity of native bird species, and should be considered seriously by VEAC and other urban land-use planners.

Future research

As systematic research into urban bird ecology is still in its relative infancy, many areas are in need of future research. We outline some of these below:

- The research described above considered the responses to urbanisation of diurnal birds of forests and woodlands. Further work is required to determine the impact of urbanisation on bird communities of other habitats such as wetlands, coastal areas and grasslands, particularly as many of these are still being cleared to make way for urban development (e.g. Williams *et al.* 2001, 2005). Although some work has been done on the distribution of nocturnal birds in Melbourne (e.g. Cooke *et al.* 2006, Isaac *et al.* 2008), further work is required;
- There is a greater need to document baseline presence/absence and relative abundance at identifiable sites, to enable changes in bird populations to be quantified over time (for example, see van Polanen Petel and Lill 2004, Platt and Lill 2006, Coates and Harris 2008). Changes could result from a number of factors, including extinction debt, increased fragmentation through loss of habitat, habitat degradation, increased or decreased competition, and climate change;
- There is a need to understand the underlying ecological mechanisms that determine the structure and composition, as well as long-term viability, of urban bird assemblages. Does the urban landscape support adequate food and breeding resources to maintain species in the longer term? For example, what is the long-term prognosis for the availability of tree hollow and the species that rely on them?
- In agricultural and forest production landscapes, there has been strong emphasis on managing and reducing the hostility of the matrix and ameliorating edge effects on vegetation remnants. Similar attention needs to be directed to understanding the role and function of 'off-reserve' vegetation in urban landscapes in promoting biodiversity, including vegetation in backyards;
- In researching aspects of introduced bird species in urban remnants, it became evident that there was a dearth of research and understanding on the ecology and impact of introduced bird species in Australia, despite general derision. For instance, the Common

Myna *Sturnus tristis* is generally considered to affect native bird species negatively as it has been shown to compete for nesting hollows (Pell and Tidemann 1997). However, it was found not to compete for food resources in Melbourne (Crisp and Lill 2006);

- A better understanding of movements and dispersal of individuals and species between urban remnants, within the urban matrix and between the urban area and beyond, is required;
- Predictions of what may happen next: if the influx of some native bird species that we are seeing today is the result of what has been planted in the 1960s-70s and the design of suburbs and reserves at the time, then what can current planning and planting tell us about the next 30 years? We need to investigate ways in which we can influence today's planning and planting to ensure that biodiversity benefits continue to increase well into the future;
- One of the great assets of cities and urban areas is the human population size – large numbers of people on hand to regularly participate in long-term surveys of urban remnants or elsewhere in the urban matrix (e.g. Birds in Backyards program). Research institutions and local governments should investigate the opportunity to harness this resource.

Acknowledgements

We thank the Parks Victoria Research Partners Panel for generously funding part of this research project. Thanks to Amber Cameron and Christina Wilson for collection of habitat data and an anonymous referee for comments on a draft of the manuscript.

References

Antos MJ, Fitzsimons JA, Palmer CG and White JG (2006) Introduced birds in urban remnant vegetation: Does remnant size really matter? *Austral Ecology* 31, 254-261.

Coates T and Harris K (2008) Birds of the Royal Botanic Gardens, Cranbourne. *Corella* 32, 1-16.

Cooke R, Wallis R, Hogan F, White J and Webster A (2006) The diet of powerful owls (*Ninox strenua*) and prey availability in a continuum of habitats from disturbed urban fringe to protected forest environments in south-eastern Australia. *Wildlife Research* 33, 199-206.

Crisp H and Lill A (2006) City slickers: habitat use and foraging in urban Common Mynas *Acridotheres tristis*. *Corella* 30, 9-15.

Daniels GD and Kirkpatrick JB (2006) Does variation in garden characteristics influence the conservation of birds in suburbia? *Biological Conservation* 133, 326-335.

Fitzsimons JA, Palmer GC, Antos MJ and White JG (2003) Refuges and residents: densities and habitat preferences of lorikeets in urban Melbourne. *Australian Field Ornithology* 20, 2-7.

Garden J, McAlpine C, Peterson A, Jones D and Possingham H (2006) Review of the ecology of Australian urban fauna: A focus on spatially explicit processes. *Austral Ecology* 31, 126-148.

Isaac B, Cooke R, Simmons D and Hogan F (2008) Predictive mapping of powerful owl (*Ninox strenua*) breeding sites using Geographical Information Systems (GIS) in urban Melbourne, Australia. *Landscape and Urban Planning* 84, 212-218.

Lowry H and Lill A (2007) Ecological factors facilitating city-dwelling in red-rumped parrots. *Wildlife Research* 34, 624-631.

Lunney D and Burgin S (eds) (2004) *Urban Wildlife: More Than Meets the Eye*. (Royal Zoological Society of New South Wales: Mosman)

Palmer GC, Fitzsimons JA, Antos MJ and White JG (2008) Determinants of native avian richness in suburban remnant vegetation: Implications for conservation planning. *Biological Conservation* 141, 2329-2341.

Parsons H (2007) *Best Practice Guidelines for Enhancing Urban Bird Habitat: Scientific Report*. Birds in Backyards Program. (Birds Australia: Melbourne) Available: <http://www.birdsinbackyards.net/documents/doc_13_guidelines_review.pdf>

Parsons H, French K and Major RE (2003) The influence of remnant bushland on the composition of suburban bird assemblages. *Landscape and Urban Planning* 66, 43-56.

Parsons H, Major RE and French K (2005) Species interactions and habitat preferences of birds inhabiting urban areas of Sydney, Australia. *Austral Ecology* 31, 217-227.

Pell AS and Tidemann CR (1997) The impact of two exotic hollow-nesting birds on two native parrots in savannah and woodland in eastern Australia. *Biological Conservation* 79, 145-153.

Platt A and Lill A (2006) Composition and conservation value of bird assemblages of urban 'habitat islands': Do pedestrian traffic and landscape variables exert an influence? *Urban Ecosystems* 9, 83-97.

Price M and Lill A (2008) Does pedestrian traffic affect the composition of 'bush bird' assemblages? *Pacific Conservation Biology* 14, 54-62.

Smith J and Lill A (2008) Importance of eucalypts in exploitation of urban parks by Rainbow and Musk Lorikeets. *Emu* 108, 187-195.

Stanford L and Lill A (2008) Out on the town: winter feeding ecology of lorikeets in urban parkland. *Corella* 32, 49-57.

van Polanen Petel T and Lill A (2004) Bird communities of some urban bushland fragments: Implications for conservation. *Australian Field Ornithology* 21, 21-32.

Weston MA, Antos MJ and Glover HK (2009) Birds, buffers and bicycles: a review and case study of wetland buffers. *The Victorian Naturalist* 126, 79-85.

White JG, Antos MJ, Fitzsimons JA and Palmer GC (2005) Non-uniform bird assemblages in urban environments: The influence of streetscape vegetation. *Landscape and Urban Planning* 71, 123-135.

Williams NSG, Leary EJ, Parris KM and McDonnell MJ (2001) The potential impact of freeways on native grassland. *The Victorian Naturalist* 118, 4-15.

Williams NSG, McDonnell MJ and Seager EJ (2005) Factors influencing the loss of an endangered ecosystem in an urbanising landscape: a case study of native grasslands from Melbourne, Australia. *Landscape and Urban Planning* 71, 35-49.

Young KM, Daniels CB and Johnston G (2007) Species of street tree is important for southern hemisphere bird trophic guilds. *Austral Ecology* 32, 541-550.

Received 27 November 2008; accepted 19 February 2009

Birds, buffers and bicycles:
a review and case study of wetland buffers

Michael A Weston[1], Mark J Antos[2] and Hayley K Glover[2]

School of Life and Environmental Sciences, Faculty of Science and Technology, Deakin University,
221 Burwood Highway, Burwood, Victoria 3125. Email: mike.weston@deakin.edu.au
Parks Victoria, Level 10, 535 Bourke Street, Melbourne, Victoria 3000.
Email: mantos@parks.vic.gov.au

Abstract
Wetland buffers separate wetlands from surrounding land uses that are incompatible with wetland values. Buffers are established to fulfil a variety of needs. However, not all functions which are attributed to buffers are mutually compatible. In particular, their use as major recreational zones is not necessarily compatible with reducing disturbance to wetland wildlife, such as birds. This paper examines the buffer around an urban wetland at Altona, Victoria, which is extensively used by recreationists. The presence of a bicycle trail within the buffer might effectively reduce its size and effectiveness, and cause 'buffer creep' whereby the effective separation distance between people and birds is reduced. It might also unintentionally facilitate unauthorised access into an otherwise 'off-limits' wetland. While social support is critical for wetland conservation, the existence of recreation in buffers does not automatically confer high awareness of local wetlands. The success of buffers as a conservation tool will depend upon setting a clear objective for buffers, careful design and management, and evaluation of effectiveness to optimise the potential benefits for wetlands and their fauna. (*The Victorian Naturalist* 126 (3) 2009, 79-86)

Keywords: Buffers, recreation, disturbance, wetlands, birds

Introduction

'Buffers' are zones that are used to separate important remnant or wetland habitat from incompatible land uses. They are used worldwide (Boyd 2001), and in Australia are a prominent feature of urban landscapes where residential and industrial development encroach on important wetland habitat (Western Australian Planning Commission (WAPC) 2005). Buffers are thought to provide benefits both to the adjacent wetland and its biodiversity, and to adjacent residents (see Table 1). In southern Australia, they are typically multiple-use zones, where a variety of human activities and management regimes are permitted or occur. Although various governments provide guidelines for minimum buffer widths (e.g. WAPC 2005), buffers vary greatly; they can be treed or grassed, actively managed (e.g. mown) or unmanaged, large or small. Although buffers are often multiuse zones, not all of the functionalities attributed to buffers are necessarily mutually compatible; for example, recreation may not be compatible with wildlife conservation (see, for example, Banks and Bryant 2007).

Despite being widely used, little is known of the effectiveness of wetland buffers in Australia (Winning 1997). In general, buffer effectiveness increases with increasing width (Castelle *et al.*

1992). However, in reality, space is at a premium in urban areas, and any land dedicated to a buffer needs to be justified. In this paper, we explore aspects of the implementation and performance of buffers from the perspective of their role in wetland wildlife conservation in urban southern Australia. We review the ways in which buffers may help conserve wildlife, and examine a case study to investigate the actual role one buffer plays in protecting an adjacent wetland of international significance to migratory shorebirds. Finally, we highlight some future research and principles that could lead to improved buffer zones.

The role of buffers in wildlife conservation

One key reason for the establishment of buffers is the protection of wildlife. Buffers may help wildlife in three direct ways:

Firstly, buffers are thought to provide additional habitat for wetland species, particularly for species that may rely on adjacent but non-wetland habitat. For example, in Massachusetts, USA, 76% of the 86 species of freshwater wetland-dependent wildlife used wetland buffers and were located at various distances from the edge of the wetland; 52% of species occurred more than 200 feet from the margin of wetlands (Boyd 2001).

Table 1. Reported functions of wetland buffers. This list builds upon functions mentioned by Anon. (1994), Winning (1997), Allan and Walker (2000), Boyd (2001) and Water and Rivers Commission (2001). Benefits are categorised into broad 'types', which may assist other workers with developing a taxonomy of benefits, which is apparently lacking at the present time.

Type of benefit	Benefit conferred to wetland or its biodiversity	Benefit conferred to adjacent residents
Wildlife and wildlife habitat	Provision of habitat and corridors for wildlife, including reducing edge effects	
	Reduced disturbance to wildlife	
	Reduced weed invasion	
	Increased public awareness of wetlands, their wildlife and threats*	
Human-centric	Improved visual amenity	Provision of recreational area
		Reduced nuisance animals
		Fire protection
Water management	Improved water quality (e.g. attenuation of pollutants, excess nutrients and sediments)	Flood mitigation
	Reduced heightened levels of runoff from surrounding areas	Erosion control
	Regulated water temperature	
	Maintenance of water levels (e.g. prevention of ground water drawdown)	
	Prevention of airborne pollutants (e.g. pesticides)	
	Accommodate for 'fuzziness' of wetland boundaries (i.e. allow for expansion in times of flood)	

* Also a benefit to adjacent residents

Secondly, buffers may provide a corridor for wildlife movement, either for wetland-dependent or terrestrial species. While the function of buffers as corridors *per se* is apparently unstudied, corridors are thought to improve connectivity between isolated habitat fragments in a landscape and to facilitate animal movement and dispersal (Beier and Noss 1998; Bennett 2003). Wetland buffers are sometimes contiguous with other wildlife habitat, especially riparian strips, and so represent an opportunity to provide a network of habitat connections between fragmented wetlands (Roe and Georges 2007).

Thirdly, buffers may reduce disturbance. 'Disturbance' is the behavioural or physiological response of an animal to a stimulus, such as a person. Documented impacts of disturbance include: displacement from habitat, such as

feeding and breeding areas; exposure of young to predators or diminished parental defence or extreme temperatures; increased conspicuousness to predators; disruption of behavioural displays, such as mating; and increased energy expenditure associated with responses (Weimerskirch *et al.* 2002; Blumstein 2003; Weston and Elgar 2005, 2007; Gill 2007).

Disturbance from human recreational activities is thought to be a key threat to some faunal groups, such as shorebirds (Burger and Gochfeld 1981; Vos *et al.* 1985; Burger and Gochfeld 1991; Fister *et al.* 1992; Weston and Elgar 2005, 2007). Buffers may reduce disturbance to wildlife in three ways:

1. A consistent finding of research into disturbance of wildlife indicates that the intensity and frequency of an animal's response is inversely proportional to the distance between the

stimulus and the animal (Cooke 1980; Rodgers and Smith 1997). Thus, by increasing the distance between people (stimuli) and animals, responses should be less frequent and less intense.

2. By facilitating the repeated presentation of benign stimuli (in this case people) to animals, buffers may underpin learning on the part of the animal whereby responses are reduced (i.e. habituation; Conomy *et al.* 1998).

3. Many buffers in southern Australia are associated with fences, and research suggests fences can decrease the impacts of human disturbance on wildlife (Ikuta and Blumstein 2003).

A Case Study: Cheetham Wetlands, Altona

In 2004 and 2005 the authors conducted a wetland conservation project, using migratory shorebirds as a flagship faunal group, at Cheetham Wetlands, south-west of Melbourne, Victoria. This study has allowed examination of some questions about the role of the buffer around wetlands with respect to wildlife conservation.

The wetlands

Cheetham Wetlands consist largely of artificial lagoons that were constructed for the commercial harvesting of salt during the 1920s (Parks Victoria 2005). The wide variety of wetland habitats available at Cheetham provides feeding, roosting and nesting areas for many species of shorebirds. The area's importance for shorebird conservation has been recognised through its listing as a wetland of international importance under the Ramsar convention (Department of Sustainability and Environment 2003). The site is home to internationally significant populations of the Sharp-tailed Sandpiper *Calidris acuminata* and Curlew Sandpiper *C. ferruginea* as well as populations of state significance of the Black-tailed Godwit *Limosa limosa*, Marsh Sandpiper *Tringa stagnatilis*, Common Greenshank *T. nebularia*, Red-necked Stint *Calidris ruficollis*, Banded Stilt *Cladorhynchus leucocephalus* and Red-necked Avocet *Recurvirostra novaehollandiae* (Watkins 1993; Lane 1997).

The wetlands are located on the western shoreline of Port Phillip Bay, only 20 km from the Melbourne CBD. As such, they are in close proximity to extensive urban development and subject to the many disturbance and degradation processes arising from those areas (Depart-

ment of Sustainability and Environment 2003). In order to maintain natural values, the wetlands are 'off-limits' to the general public, and a buffer, which hosts a bicycle trail, is maintained between the residences and the wetlands. Parks Victoria (1997), the manager of the wetlands, stated 'a strip of land around the perimeter ... has been identified as a buffer to the environmentally sensitive area. It is proposed that the Bay [bicycle] Trail will be located in this area'. The interface between an area of high natural values and extensive residential development means Cheetham Wetlands are an ideal model for examining wetland buffers.

Buffer creep

The term 'buffer creep' is used to describe the circumstance whereby the effective separation distance between incompatible activities and a wetland is unintentionally shifted in space, while the physical extent of the area designated as the buffer remains the same (see Fig. 1). At Cheetham Wetlands, a sealed, formal bicycle track now runs the length of the buffer, and so the effective separation distance between recreationists and wildlife is decreased; a track down the middle of a buffer would halve the effective buffer distance in terms of protecting wildlife from human disturbance. If it is assumed that there is a consistent tolerance distance of wildlife to humans (Cooke 1980; Rodgers and Smith 1995, 1997), then the effective separation distance has been shifted ('crept') into the wetland. We have no data on whether buffer creep is evident at Cheetham Wetlands or elsewhere, and such studies would be instructive.

Buffer creep may be especially evident where human presence is highly concentrated in space, such as by a formed bicycle track. In Melbourne and many cities around the world, such trails are extensive and expanding, often following watercourses and coastlines (Dill and Carr 2003). At Cheetham Wetlands, the vast majority of recreationists occurred on the bicycle trail, but a substantial minority occurred off the trail on adjacent grassed areas, including some who walked dogs on the side of the buffer nearest the wetlands (pers. obs.).

Do recreationists use the buffer?

As part of our general study of the Cheetham Wetlands (Antos *et al.* 2007), we conducted six hours of observations, in three two-hour blocks, for each of four Sundays (summer

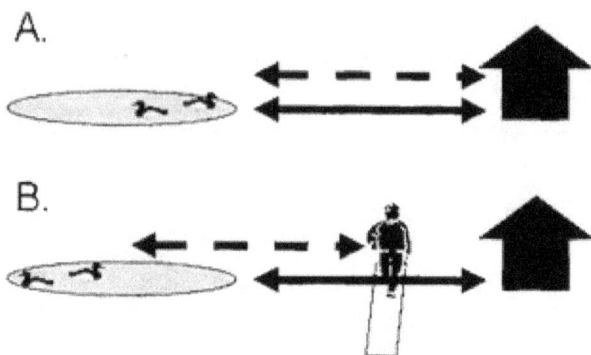

Fig. 1. An illustration of 'buffer creep'. The figure presents two scenarios, a buffer without incompatible activities (A) and a buffer with a recreational path (B). Width of the physical buffer (solid arrows) remains unchanged between scenarios while the effective separation distance between the wetland and incompatible activities (dashed arrow) effectively shifts with the introduction of a recreational path into the buffer. Under scenario B, the buffer has shrunk, but because wildlife response distances probably remain constant, the effective buffer now extends into the wetland.

2004/2005), from a vantage point (37°53'01"S, 144°47'50"E) that enabled a clear view of 1.7 km of the bicycle trail within the buffer to the south west. The furthest point which we could see was where the trail joined the Skeleton Creek trail (37°53'41"S, 144°46'59"E). Binoculars and spotting scopes were used to obtain clear views of all recreationists (refer to Antos *et al.* 2007 for a more detailed description of the site and a site map).

Recreationists used the trail in each of the twelve observation periods (Fig. 2). Overall, 25.6±9.1 (sd) people used the trail each hour. Nearly half (43%) of all recreationists within the buffer were cycling, and a range of other recreational activities also occurred (Fig. 2). These results demonstrate that the buffer is used extensively by recreationists for a variety of activities.

Do recreationists in buffers obey regulations?
Dogs must be leashed in the buffer; however, 68.3% of dogs observed (n = 104) were unleashed. Unleashed dogs are particularly disturbing to birds, including shorebirds (Banks and Bryant 2007; Weston and Elgar 2007). No evidence of dog regulation enforcement was observed during observations.

Trail bikes (off-road motorcycles) are not permitted on the trail, and the local police actively patrolled the bicycle trail with a view to curbing this illegal activity (pers. obs.). Nevertheless,

trail bikes occurred on the trail (Fig. 2). Moving rapidly and being noisy, motorised transportation can be potentially highly disturbing to wildlife (see Garcia and Baldassare 2008). It also raises obvious safety concerns for other users of the bicycle path.

The level of compliance may vary in relation to location, education and enforcement activities (Gramann and Bonifield 1995; Solomon 1998; Kasapoglu and Ecevit 2002) but observations suggest many recreationists in buffers do not obey regulations intended to help reduce disturbance to wildlife. The buffer currently has interpretive and regulatory signs, and a variety of education and extension programs have been conducted in the vicinity of the wetlands (Antos *et al.* 2007).

Do buffers facilitate intrusions rather than prevent them?
Currently, the northern half of Cheetham Wetlands has a recreational path through the buffer, and construction of the path through the remainder of the buffer along the southern half of the wetlands has commenced. Antos *et al.* (2007) found that virtually all unauthorised human intrusions into the wetland occurred in the northern half, where the buffer and path bound the wetland. This suggests that the path might facilitate unauthorised access, a contention supported by a number of well-established informal paths leading from the recreational

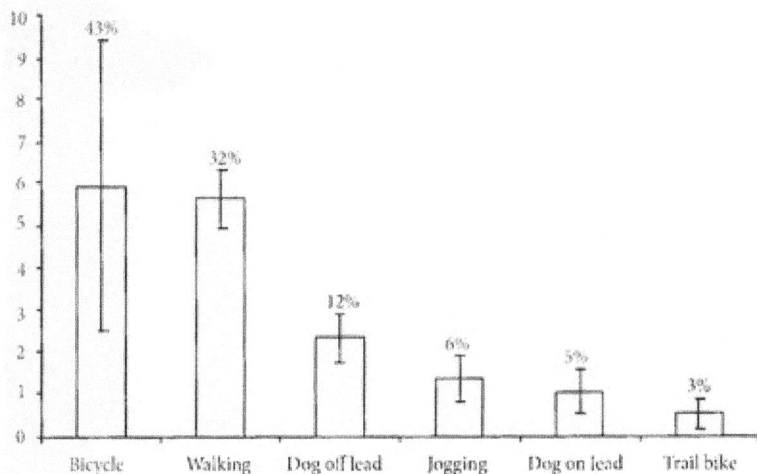

Fig. 2. Mean (± one standard error) number of groups of humans engaged in different recreational pursuits. The figures above the bars indicate the percentage of all humans (n = 614) engaged in different activities.

path, through the boundary fence and into the wetlands (Fig. 3). However, this assertion should be treated with caution because the northern and southern half of Cheetham Wetlands differ in other respects, such as the southern half not currently having abutting residential development.

Does recreation in buffers improve awareness of wetlands?

Human appreciation and understanding of wetlands is crucial to their conservation (Shumula 2002; Bouton and Frederick 2003). Education and awareness play key roles in developing attitudes and appreciation of important habitats like wetlands (McKenzie-Mohr and Smith 1999). It would be interesting to know whether allowing people to access the margins of Cheetham Wetlands had raised their levels of awareness and appreciation of the wetlands. Weston et al. (2006) surveyed primary school students at a local school to examine awareness levels of wetlands around Cheetham. They found local wetlands and parks varied dramatically in respect of how well known they were (0-91%). Surprisingly, no students reported awareness of the Cheetham Wetlands despite the fact they were only 200 m away from their

school. Most students displayed moderately positive attitudes to wetlands and wetland values. While this study did not directly examine the role that recreational opportunities in the buffer played in awareness among students, it suggests that the presence of recreational opportunities in a buffer does not automatically confer high awareness of significant wetlands.

Towards better buffers

Buffers have the potential to provide protection for wetlands and their biodiversity from adjoining land uses, provided they are well-planned and appropriately managed. However, their performance is little studied, especially in view of their multiple functions (Winning 1997). We suggest that two steps could improve the effectiveness of buffers:

1. Higher specificity of management goals of buffers would aid their design and implementation, and avoid unwanted generality or ambiguity with respect to their objectives (Castelle et al. 1992). Specifically, we suggest the proposed role of the buffer in the conservation of wildlife should be stated explicitly as one or a combination of: (a) provision of habitat, (b) provision of corridors and/or (c) reduction of disturbance. Each goal potentially engenders

Fig. 3. A well-established unauthorised path leading from the bicycle trail, through the boundary fence into Cheetham Wetlands, Altona, Victoria. Photo by MA Weston.

different buffer designs, management, and balance between recreational and wildlife needs. The management of buffers should reflect their identified roles.

2. Research that addresses key questions about buffer design and management is needed. The optimal design of buffers intended to provide habitat and corridors would usefully draw from the body of research on landscape ecology and reserve system design (e.g. Beier and Noss 1998; Cabeza and Moilanen 2001). Buffers to minimise disturbance could utilise Flight Initiation Distances, which are currently available from overseas (Blumstein *et al.* 2003), but are largely unavailable for Australian wetland birds. The determination of buffer widths should also account for the specific objective of a buffer (Castelle *et al.* 1992; Allan and Walker 2000) and for fluctuating water levels (WAPC 2005).

3. The creation of ecologically meaningful guidelines for the establishment of buffers is imperative if they are to fulfil the role of enhancing nature conservation. Such guidelines should be informed by appropriate science, much of which is not yet available, especially in the Australian

context. Students of ecology, conservation biology and environmental management are encouraged to better investigate the strengths, weaknesses and opportunities that buffers present for the conservation of wildlife and habitat, by conducting studies that provide results that are readily available for use by planners and managers. The monitoring of the effectiveness of established buffers and a willingness to engage in adaptive management to ensure they fulfil their designated roles is also desirable.

Acknowledgments
The Cheetham Wetlands project, including all data collection, was conducted by Birds Australia, and funded by the Australian Government's Natural Heritage Trust and Price Waterhouse Coopers, and managed by WWF Australia. We thank the Parks Victoria rangers at the Cheetham Wetlands, especially Bernie McCarrick and John Argote, for their time and generous assistance with our research. Bianca Priest, Glenn Ehmke and Chris Tzaros all provided valuable input into the Cheetham Wetlands Project for which we are grateful. This paper was given at the FNCV Biodiversity Symposium on urban birds (September 2008), and emphasises the need for buffers for wetland birds in the Melbourne area. It intends no criticism of management at Cheetham Wetlands.

References

Allan M and Walker C (2000) *Wetland Buffers. Advisory Notes for Land Managers on River and Wetland Restoration.* (WA Government: Perth)

Anon. (1994) *Tidal Wetland Buffer Guidance Document.* (Long Island Sounds Programs: Connecticut, USA)

Antos MJ, Ehmke GC, Tzaros CL and Weston MA (2007) Unauthorised human use of an urban coastal wetland sanctuary: Current and future patterns. *Landscape and Urban Planning* 80, 173–183.

Banks PV and Bryant JB (2007) Four-legged friend or foe? Dog walking displaces native birds from natural areas. *Biology Letters* 3, 611–613.

Beier P and Noss RF (1998) Do habitat corridors provide connectivity? *Conservation Biology* 12, 1241-1252.

Bennett A (2003). *Linkages in the Landscape: The Role of Corridors and Connectivity in Wildlife Conservation.* (IUCN: Gland, Switzerland)

Blumstein DT (2003) Flight-initiation distance in birds is dependent on intruder starting distance. *The Journal of Wildlife Management* 67, 852-857.

Blumstein DT, Anthony LL, Harcourt R and Ross G (2003) Testing a key assumption of wildlife buffer zones: is flight initiation distance a species-specific trait? *Biological Conservation* 110, 97–100.

Bouton SN and Frederick PC (2003) Stakeholders' perceptions of a wading bird colony as a community resource in the Brazilian Pantanal. *Conservation Biology* 17, 297-306.

Boyd L (2001) *Buffer zones and beyond. Wildlife use of wetland buffer zones and their protection under the Massachusetts Wetland Protection Act.* (Wetland Conservation Professional Program Department of Natural Resources Conservation, University of Massachusetts: USA)

Burger J and Gochfeld M (1981) Discrimination of the threat of direct versus tangential approach to the nest by incubating Herring and Great Black-backed Gulls. *Journal of Comparative Physiology and Psychology* 95, 676-684.

Burger, J and Gochfeld M (1991) Human activity influence and diurnal and nocturnal foraging of sanderlings (*Calidris alba*). *The Condor* 93, 259-265.

Cabeza M and Moilanen A (2001) Design of reserve networks and the persistence of biodiversity. *Trends in Ecology and Evolution* 16, 242-248.

Castelle AJ, Conolly C, Emers M, Metz ED, Meyer S, Witter M, Mauermann S, Erickson T and Cooke SS (1992) *Wetland buffers: use and effectiveness.* Adolfson Associates, Inc., Shorelands and Coastal Zone Management Program, Washington Department of Ecology, Olympia, Pub. No. 92-10.

Conomy JT, Dubovsky JA, Collazo JA and Fleming WJ (1998) Do Black Ducks and Wood Ducks habituate to aircraft disturbance? *The Journal of Wildlife Management* 62, 1135–1142.

Cooke AS (1980) Observations on how close certain passerine species will tolerate an approaching human in rural and suburban areas. *Biological Conservation* 18, 85-88.

Department of Sustainability and Environment (2003) *Port Phillip Bay (Western Shoreline) and Bellarine Peninsula Ramsar Site Strategic Management Plan.* (DSE: Melbourne)

Dill J and Carr T (2003) Bicycle commuting and facilities in major U.S. cities: If you build them, commuters will use them – another look. Proceedings of the Transport Research Bureau 2003 Annual Meeting, USA.

Fister C, Harrington BA and Lavine DM (1992) The impact of human disturbance on shorebirds at a migration staging area. *Biological Conservation* 60, 115-126.

Garcia E and Baldassare GA (2008) Effects of motorized tourboats on the behavior of non-breeding American Flamingos in Yucatan, Mexico. *Conservation Biology* 11, 1159-1165.

Gill J (2007) Approaches to measuring the effects of human disturbance on birds. *Ibis* 149, 9-14.

Gramann JH and Bonifield RL (1995) Effect of personality and situational factors on intentions to obey rules in outdoor recreation areas. *Journal of Leisure Research* 27, 326-343.

Ikuta LA and Blumstein DT (2003) Do fences protect birds from human disturbance? *Biological Conservation* 112, 447-452.

Kasapoglu MA and Ecevit MC (2002) Attitudes and behaviour toward the environment. The case of Lake Burdur in Turkey. *Environment and Behaviour* 34, 363-377.

Lane BA (1997) Monitoring program for determining the impact of human visitation on avifauna at the Cheetham Wetlands (Ecology Australia: Fairfield)

McKenzie-Mohr D and Smith W (1999) *Fostering Sustainable Behaviour: An Introduction to Community-based Social Marketing* (New Society Publishers: Gabriola Island, USA).

Parks Victoria (1997) *Point Cook Coastal Park and Cheetham Wetlands Strategy Plan* (Parks Victoria: Melbourne)

Parks Victoria (2005) *Point Cook Coastal Park and Cheetham Wetlands Future Directions Plan* (Parks Victoria: Melbourne)

Rodgers JA and Smith HT (1995) Set-back distances to protect nesting bird colonies from human disturbance in Florida. *Conservation Biology* 9, 89-99.

Rodgers JA and Smith HT (1997) Buffer zone distances to protect foraging and loafing waterbirds from human disturbance in Florida. *Wildlife Society Bulletin* 25, 139-145.

Roe JH and Georges A (2007) Heterogeneous wetland complexes, buffer zones, and travel corridors: Landscape management for freshwater reptiles. *Biological Conservation* 135, 67–76.

Shunula JP (2002) Public awareness, key to mangrove management and conservation: the case of Zanzibar. *Trees – Structure and Function* 16, 209-212.

Solomon BD (1998) Public support for endangered species recovery: an exploratory study of the Kirtland's Warbler. *Human Dimensions of Wildlife* 3, 62-74.

Vos DK, Ryder RA and Graul WD (1985) Response of breeding Great Blue Herons to human disturbance in north-central Colorado. *Colonial Waterbirds* 8, 13-22.

Watkins D (1993) A national plan for shorebird conservation in Australia. (Australasian Wader Studies Group, Royal Australasian Ornithologists Union and World Wide Fund for Nature: Melbourne)

Weimerskirch H, Shaffer SA, Mabille G, Martin J, Boutard O and Rouanet JL (2002) Heart rate and energy expenditure of incubating Wandering Albatrosses: basal levels, natural variation, and the effects of human disturbance. *The Journal of Experimental Biology* 205, 475-483.

Western Australian Planning Commission (WAPC) (2005) *Guideline for the Determination of Wetland Buffer Requirements.* (Department for Planning and Infrastructure, Western Australian Planning Commission and Essential Environmental Services: Perth)

Weston MA and Elgar MA (2005) Disturbance to broodrearing Hooded Plover *Thinornis rubricollis*: responses and consequences. *Bird Conservation International* 15, 193–299.

Weston MA and Elgar MA (2007) Responses of incubating Hooded Plover (*Thinornis rubricollis*) to disturbance. *Journal of Coastal Research* 23, 569 – 576.

Weston MA, Tzaros CL and Antos MJ (2006) Awareness of wetlands and their conservation value among students at a primary school in Victoria, Australia. *Ecological Management and Restoration* 7, 223-226.

Winning G (1997) The functions and widths of wetland buffers. Hunter Wetlands Research Technical Memorandum 1.

Water and Rivers Commission (WRC) (2001) Position statement on wetlands. (WA Government: Perth)

Received 20 November 2008, accepted 12 March 2009

Red-necked stints *Callidris ruficollis*. Photo by MA Weston

The presence of a human causes the shorebirds to flush. Photo by MA Weston

Feral Mallards: a risk for hybridisation with wild Pacific Black Ducks in Australia?

P-J Guay[1,2] and JP Tracey[1]

[1]Department of Zoology, The University of Melbourne, Victoria 3010
[2]Current address: Institute for Sustainability and Innovation, Victoria University,
St. Albans Campus, P O Box 14428C, MCMC, Victoria 8001
Invasive Animals Cooperative Research Centre, Vertebrate Pest Research Unit, NSW Department of
Primary Industries, Forest Rd, Orange, NSW, Australia

Abstract

Hybridisation is widespread in waterfowl and hybrids are often fertile. Mallards *Anas platyrhynchos* hybridise with numerous dabbling ducks and have been associated with decline in many *Anas* species with which they co-occur. Mallards have been introduced in Australia and New Zealand where they hybridise with indigenous Pacific Black Ducks *Anas superciliosa*. The extent of hybridisation in Australia is unknown, but Mallards pose a potentially serious threat to endemic duck populations and have already caused the extinction of some populations of Pacific Black Ducks in New Zealand, Lord Howe Island and Macquarie Island. The distribution and abundance of Mallards and the extent of hybridisation in Australia must be determined as a priority to ensure the long-term genetic integrity of the Pacific Black Duck. (*The Victorian Naturalist* 126 (3) 2009, 87-91).

Keywords: Pacific Black Duck, Mallard, Hybridisation, Impacts, Australia

Introduction

The Pacific Black Duck *Anas superciliosa* is a dabbling duck native to the West Pacific, and is sometimes divided into three subspecies: the Australian Black Duck *A. s. rogersi* from Australia, New Guinea and Indonesia, the New Zealand Grey Duck *A. s. superciliosa* from New Zealand and outlying islands, and the Lesser Grey Duck *A. s. pelewensis* from Pelew and other islands in the West Pacific (Amadon 1943; Rhymer *et al.* 2004). The Lesser Grey Duck is much smaller and easily distinguished from the two other subspecies (Frith 1967; Schodde 1977). The New Zealand Grey Duck and the Australian Black Duck are of similar size and may differ only marginally in plumage (Frith 1967). Both the New Zealand Grey Duck and the Lesser Grey Duck are declining (Threatened Waterfowl Specialist Group 2003). In both cases, hybridisation with either wild or domestic Mallard *A. platyrhynchos* is likely to play a role in reducing their genetic integrity and may contribute to significant population declines.

Hybridisation in waterfowl

Hybridisation in waterfowl is quite common, with more than 500 hybrid combinations reported for Anseriformes in the wild or in captivity (McCarthy 2006). Interestingly, many of those hybrids are fertile, while in other orders chromosomal imbalances cause hybrid sterility (McCarthy 2006). Mallards are known to hybridise with 40 species in the wild and a further 20 species in captivity (McCarthy 2006). Seventeen of these naturally occurring crosses produce hybrids that are at least partially fertile (McCarthy 2006). In extreme cases, hybridisation can lead to introgression, the mixing of the gene pool of the two species that hybridise to such a degree that the individual gene pool of the two parents does not exist anymore, which can lead to extinction if left unchecked (Rhymer 2006). This has occurred on the West Pacific Islands of Guam, Tinian and Saipan where Mallards and Pacific Black Duck hybridised to such an extent that both species became locally extinct and only hybrids were left (Yamashina 1948).

Hybridisation and introgression with Mallards is thought to be one of the important threats causing the decrease in some endangered species, the Hawaiian Duck *A. wyvilliana* (Engilis *et al.* 2002), the Meller's Duck *A. melleri* (Jones 1996) and the New Zealand Grey Duck *A. s. superciliosa* (Gillespie 1985), and has been found to threaten three other species, the Mexican Duck *A. diazi* (Short 1978), the American Black Duck *Anas rubripes* (Ankney *et al.* 1987) and the Yellow-billed Duck *A. undulata* (Milstein and Osterhoff 1975). Mallards are

physically dominant over, and have been impli
cated as a threat to, Pacific Black Ducks wherev-
er they co-occur, due to their increased capacity
for survival, greater fecundity and productivity,
and willingness to exploit human environments
(Williams and Basse 2006). Thus, Mallards hybridise with and threaten Pacific Black Ducks
in New Zealand (Gillespie 1985), Lord Howe
Island (Tracey *et al.* 2008), Macquarie Island
(Norman 1987) and the Australian mainland
(Braithwaite and Miller 1975).

New Zealand

Pacific Black Ducks are thought to have colonised New Zealand from Australia somewhere
between 5000 and 20000 years ago (Hitchmough *et al.* 1990). Mallards were first introduced on the South Island from European
game-farm stocks by the Otago Acclimatiza-
tion Society in the mid-1800s, and later on,
birds from North America were introduced on
the North Island by the Auckland Acclimatiza-
tion Society in the 1930s (Williams 1981). By
1963 over 20000 Mallards had been released
throughout New Zealand (McDowall 1994).
In the early 1980s the Mallard population had
reached 5 million and was still increasing,
while the Pacific Black Duck population had
decreased to 1.5 million and was still declining
(Marchant and Higgins 1990). The first hybrid
was reported in 1917 and since then hybrid frequency has increased steadily (Thomson 1922;
Gillespie 1985). By the early 1980s, hybrid frequency was estimated at 51% and pure Pacific
Black Duck frequency at 4.5% (Gillespie 1985).
This is below 5%, the minimum level that is
suggested to ensure species survival (Short
1969). The current extent of hybridisation is
unknown, but after 24 years since the study by
Gillespie (1985), most populations of Pacific
Black Duck in New Zealand are now likely to
have had some level of exposure to Mallards.
Hybrid phenotypes are not continuous in New
Zealand, but two types of hybrid exist, Pacific Black Duck-like and Mallard-like hybrids
(Gillespie 1985). This is similar to the situation
that was observed in the Mariana Mallard *Anas
oustaleti* that is thought to have originated from
hybrids between Mallards and Pacific Black
Duck in the Western Pacific Islands of Guam,
Tinian and Saipan (Yamashina 1948; Reichel
and Lemke 1994).

Lord Howe Island

Lord Howe Island (159°05' E, 31°33' S) is situated at the intersection of the range of the three
Pacific Black Duck subspecies in the Pacific
Ocean east of Australia. Pacific Black Ducks
were first recorded on Lord Howe Island in
1852 and breeding has been observed, albeit in-
frequently, ever since (MacDonald 1853; Hindwood and Cunningham 1950; Rogers 1972).
Mallards were first observed in 1963 and probably self-introduced from New Zealand (McK-
ean and Hindwood 1965; Tracey *et al.* 2008).
Hybridisation with the local Pacific Black Duck
population followed soon after (Rogers 1976).
Recent surveys suggest that hybridisation has
progressed and that pure Pacific Black Ducks
are now absent (Tracey *et al.* 2008). Furthermore, Pacific Black Duck-like hybrids represent
only 2% (*s.e.* = 0.59, *n* = 86) of the population
(Tracey *et al.* 2008). Pacific Black Ducks are
thus effectively extinct on the island and only
extermination of Mallards and their hybrids
would allow Pacific Black Ducks to re-establish
themselves (Tracey *et al.* 2008).

Macquarie Island

Macquarie Island (158°53' E, 54°35' S) is situated south-east of Tasmania and south-west of
New Zealand (Copson 1984; Norman 1987).
The island has a small resident breeding population of Pacific Black Ducks from at least the
late 1800s (Hamilton 1895, Norman 1987). Mallards were first recorded on the island in 1949
(Gwynn 1953). During the following 25 years, a
few scattered records of Mallards exist, but they
do not seem to have established themselves
until 1975 (Norman 1987). The provenance of
the Mallards has not been established, but they
probably self-introduced from New Zealand
or maybe from Campbell or Auckland Islands
(Norman 1987, 1990, OSNZ 1990). A Mallard
x Pacific Black Duck hybrid was observed in
1973, but it is unclear if it had been bred locally
(Norman 1987). Hybrids became common in
1978 and by 1985 most ducks observed on the
island were hybrids (Norman 1987, 1990).

Australian Mainland

Mallards were introduced to Australia for recreational hunting from the 1860s, but unlike in
New Zealand, they did not colonise the whole
country and were restricted to areas around
urban centres (Marchant and Higgins 1990).

Most of the Mallards in Australia today are of domestic origin. Because of their domestic origin, Mallards are 50% heavier than Pacific Black Ducks and have highly variable plumage (Marchant and Higgins 1990). The feral Mallards are thought to rely on urban habitat as they were previously not recorded in hunter bag surveys (Braithwaite and Norman 1974). However, recent anecdotal information from hunters suggests hybrids are occurring more frequently in rural areas (B. Boyle, Game Council NSW, pers. comm.; P.-J. Guay unpublished data). Pacific Black Ducks have a much wider distribution, being present throughout Australia except for the central desert (Marchant and Higgins 1990). Mallards and Pacific Black Ducks hybridise extensively and their progeny have been reported from various parts of the country (Braithwaite and Miller 1975; Whatmough 1978; Smith and Smith 1990; Paton *et al.* 1992; Bielewicz and Bielewicz 1996). Neither Mallards nor hybrids are reported in hunter bag surveys (Braithwaite and Norman 1974, 1976), but both occur in urban wetlands. This led Braithwaite and Miller (1975) to conclude that hybridisation is limited to cities and is not a threat to Pacific Black Ducks in Australia. Since then, concerns about the future of Pacific Black Ducks in the presence of Mallards have been expressed by a number of authors (Parker *et al.* 1985; Paton *et al.* 1992; Rhymer *et al.* 2004; Williams and Basse 2006). The lack of hybrids in hunter bags may suggest that, like Mallards, F1 hybrids are mostly sedentary in urban wetlands and that offspring of back crosses with Pacific Black Ducks disperse out of cities. Phenotypical studies have demonstrated that hybridisation is spreading in South Australia (Paton *et al.* 1992) and that hybrids are now shot by hunters in rural parts of Tasmania, New South Wales and Victoria (B. Boyle, Game Council NSW, personal communication, P.-J. Guay unpublished data). Investigations limited to phenotypes may also be underestimating the extent of the problem in Australia, as first generation hybrids may look like Pacific Black Ducks as the white plumage of some breeds of domestic Mallards is likely to be recessive to the wild-type Pacific Black Duck plumage.

Dynamics of Mallard x Pacific Black Duck hybridisation

Mallards and Pacific Black Ducks have similar courtship behaviour (Lorenz 1951; Williams 1969), but pair assortively in New Zealand (Gillespie 1985; Hitchmough *et al.* 1990). Similarly, domestic Mallards, Pacific Black Ducks and their hybrids pair assortively on urban wetlands in Australia (Braithwaite and Miller 1975). Courtship behaviour can be quite degenerate in domestic breeds compared to wild Mallard (Desforges and Wood-Gush 1976; but see Miller 1977), which could account for lack of pair formation between domestic Mallards and Pacific Black Ducks (Braithwaite and Miller 1975). Mallards are very aggressive and will dominate most species and can hybridise through forced copulation even if no interspecific pairing occurs (Ankney *et al.* 1987; Brodsky *et al.* 1988; Hitchmough *et al.* 1990). Forced copulations are an important aspect of Mallards mating strategy and can lead to egg fertilisation (Burns *et al.* 1980; McKinney and Evarts 1997). Furthermore, domestication in Mallards has increased reliance on forced copulation to secure mating opportunities (Desforges and Wood-Gush 1976; but see Miller 1977). Simple monitoring of pairing on wetlands where both Mallards and Pacific Black Ducks co-occur may not tell the whole story.

Dispersal of hybrids

Pacific Black Ducks are highly dispersive (Marchant and Higgins 1990). Band recoveries have demonstrated that they can disperse over the Nullarbor Plain, Bass Strait and the Tasman Sea (Braithwaite and Miller 1975; Marchant and Higgins 1990; Halse *et al.* 2005). In contrast, Mallards seem to be mostly sedentary in Australia and New Zealand (Marchant and Higgins 1990). At fledging, F1 hybrids or F2 back crosses may disperse outside of urban wetlands with other Pacific Black Ducks, spreading the Mallard genes into wild populations.

Monitoring options

Both the phenotype and the genotype of hybrids have been studied (e.g. Rhymer *et al.* 1994). Various quantitative and qualitative hybrid phenotype classification schemes relying on plumage and bare parts have been established (Braithwaite and Miller 1975; Gillespie 1985; Paton *et al.* 1992; Rhymer *et al.* 1994). These have been used to evaluate the extent of hybridisation in various populations (e.g. Tracey *et al.* 2008). The use of phenotype may be limited because, in hybrid swarms, phenotypes tend to be bimodally distributed and to converge toward both parental types (e.g. Yamashina 1948), thereby complicating analyses based solely on phenotypes. Furthermore, it is difficult to differentiate hybrids from parental species phenotypically after a few generations of backcrossing (Rhymer *et al.* 1994, Green *et al.* 2000). Alternatively, hybridisation can be monitored genetically: Mallards and Pacific Black Ducks have distinct mitochondrial control region sequences (Rhymer *et al.* 1994). Mitochondrial haplotypes only allow the determination of maternal lines because mitochondrial genes are exclusively maternally inherited (Watanabe *et al.* 1985). This may not be a problem for hybridisation between Mallards and Pacific Black Ducks because both sexes mate interspecifically in both species. Nuclear genes in contrast would allow better quantitation of hybridisation since they are inherited from both parents (e.g. Fowler *et al.* In Press). Unfortunately, no such markers are currently available for Pacific Black Ducks and Mallards.

Conservation implications

The extent of hybridisation between Mallards and Pacific Black Ducks in Australia is still poorly understood. While this is often ignored or dismissed, a major potential threat exists. Mallards and their hybrids are known to cause the extinction of Pacific Black Duck populations, as demonstrated in New Zealand, Lord Howe Island and Macquarie Island, and they are increasingly being recorded in rural Australia. To assess the associated risks to Pacific Black Duck, the distribution and abundance of Mallards and the extent of hybridisation must be determined as a priority. New nuclear markers must be developed to allow genotyping of ducks and determination of extent of hybridisation in wild populations. Appropriate management techniques can then be employed to control Mallards and their hybrids both in urban and rural settings. Continued control efforts may also be necessary as wild Mallards are known to travel from New Zealand to Australia (Paton 1991) and further self-introduction of wild Mallards from New Zealand to Australia is an additional on-going threat to the genetic integrity of the Pacific Black Duck.

Acknowledgements
Funding for this work was provided by Museum Victoria, The Game Management Unit from the Tasmanian Department of Primary Industries and Water, the Invasive Animals Cooperative Research Centre and the Vertebrate Pest Research Unit of the New South Wales Department of Primary Industries.

References
Amadon D (1943) Birds collected during the Whitney south sea expedition. *American Museum Novitates* **1237**, 1-22.
Ankney CD, Dennis DG and Bailey RC (1987) Increasing Mallards, decreasing American Black Ducks: coincidence or cause and effect? *Journal of Wildlife Management* **51**, 523-529.
Bielewicz J and Bielewicz F (1996) The birds of the Redcliffe Peninsula, southeast Queensland. *Sunbird* **26**, 81-120.
Braithwaite LW and Miller B (1975) The Mallard, *Anas platyrhynchos*, and the Mallard-Black Duck, *Anas superciliosa rogersi*, hybridization. *Australian Wildlife Research* **2**, 47-61.
Braithwaite LW and Norman FI (1974) The 1972 open season on waterfowl in south-eastern Australia. *CSIRO Division of Wildlife Research Technical Paper* **29**, 1-37.
Braithwaite LW and Norman FI (1976) The 1973 and 1974 open seasons on waterfowl in south-eastern Australia. *CSIRO Division of Wildlife Research Technical Memorandum* **11**, 1-58.
Brodsky LM, Ankney CD and Dennis DG (1988) The influence of male dominance on social interactions in Black Ducks and Mallards. *Animal Behaviour* **86**, 1371-1378.
Burns JT, Cheng KM and McKinney F (1980) Forced copulation in captive Mallards I. Fertilization of eggs. *Auk* **97**, 875-879.
Copson GR (1984) An annotated atlas of the vascular flora of Macquarie Island. *ANARE Research Notes* **18**.
Desforges MF and Wood-Gush DGM (1976) Behavioural comparison of Aylesbury and Mallard ducks: Sexual behaviour. *Animal Behaviour* **24**, 391-397.
Engilis AJ, Uyehara KJ and Griffin JG (2002) Hawaiian Duck (*Anas wyvilliana*). In *The Birds of North America, No. 694*. Eds A Poole and F Gill (The Birds of North America, Inc: Philadelphia, PA)
Fowler AC, Fadie JM and Engilis AJ (in press) Identification of endangered Hawaiian Ducks (*Anas wyvilliana*) introduced North American Mallards (*A. platyrhynchos*) and their hybrids using multilocus genotypes. *Conservation Genetics*.
Frith HJ (1967) *Waterfowl in Australia*. (Angus and Robertson Ltd: Sydney, Australia)
Gillespie GD (1985) Hybridization, introgression, and morphometric differentiation between Mallards (Anas platyrhynchos) and Grey Ducks (*Anas superciliosa*) in Otago, New Zealand. *Auk* **102**, 459-469.
Green J, Wallis G and Williams M (2000) *DOC Science Poster No. 32: Determining the extent of Grey Duck X Mallard hybridisation in New Zealand*. (Department of Conservation: Wellington, New Zealand)
Gwynn AM (1953) Some additions to the Macquarie Island list of birds. *Emu* **53**, 150-152.
Halse SA, Pearson GB, Hassell C, Collins P, Scanlon MD and Minton CDT (2005) Mandora Marsh, north-Western Australia, an arid-zone wetland maintaining continental populations of waterbirds. *Emu* **105**, 115-125.
Hamilton A (1895) Notes on a visit to Macquarie Island. *Transactions and Proceedings of the New Zealand Institute* **27**, 559-579.
Hindwood KA and Cunningham JM (1950) Notes on the birds of Lord Howe Island. *Emu* **50**, 23-35.
Hitchmough RA, Williams M and Daugherty CH (1990) A genetic analysis of Mallards, Grey Ducks, and their hybrids in New Zealand. *New Zealand Journal of Zoology* **17**, 467-472.
Jones CG (1996) Bird introductions to Mauritius: Status and relationships with native birds. In *The introduction and naturalisation of birds*, pp. 113-123. Eds JS Holmes and JR Simons (Her Majesty's Stationery Office, London, UK)
Lorenz KL (1951) Comparative studies on the behaviour of the Anatidae. *Avicultural Magazine* **57**, 157-182.
MacDonald JD (1853) Remarks on the natural history and

capabilities of Lord Howe Island. In *Proposed Penal Settlement*, pp. 13-17. (Legislative Council: Sydney, Australia.)

Marchant S and Higgins PJ (1990) *Handbook of Australian, New Zealand, and Antarctic Birds. Vol. 1B Pelican to Ducks.* (Oxford University Press: Oxford)

McCarthy EM (2006) *Handbook of Avian Hybrids of the World.* (Oxford University Press: Oxford)

McDowall RW (1994) *Gamekeepers for the Nation.* (Canterbury University Press: Christchurch, New Zealand)

McKean JL and Hindwood KA (1965) Additional notes on the birds of Lord Howe Island. *Emu* 64, 79-97.

McKinney F and Evarts S (1997) Sexual coercion in waterfowl and other birds. *Ornithological Monographs* 49, 163-195.

Miller DB (1977) Social displays of Mallard Ducks (*Anas platyrhynchos*) - Effects of domestication. *Journal of Comparative and Physiological Psychology* 91, 221-232.

Milstein PI and Osterhoff DR (1975) Blood-tests identify by brid wild ducks new to science. *Bokmakierie* 27, 70-71.

Norman H (1987) The ducks of Macquarie Island. *ANARE Research Notes* 42.

Norman H (1990) Macquarie Island ducks - Habitats and hybrids. *Notornis* 37, 53-58.

OSNZ (1990) *Checklist of the Birds of New Zealand and the Ross Dependency, Antarctica.* (Random Century New Zealand Ltd: Auckland, New Zealand)

Parker SA, Eckert HJ and Ragless GB (1985) *An Annotated Checklist of the Birds of South Australia, Part 2A: Waterfowl.* (The South Australian Ornithological Association: Adelaide, Australia)

Paton JB (1991) Movement of a Mallard from New Zealand to South Australia. *South Australian Ornithologist* 31, 73.

Paton JB, Storr R, Delroy L and Best L (1992) Patterns to the distribution and abundance of Mallards, Pacific Black Ducks and their hybrids in South Australia in 1987. *South Australian Ornithologist* 31, 103-110.

Reichel JD and Lemke TO (1994) Ecology and extinction of the Marianna Mallard. *Journal of Wildlife Management* 58, 199-205.

Rhymer JM (2006) S33-4 Extinction by hybridization and introgression in anatine ducks. *Acta Zoologica Sinica* 52 (Suppl), 583-585.

Rhymer JM, Williams MJ and Braun MJ (1994) Mitochondrial analysis of gene flow between New Zealand Mallards (*Anas platyrhynchos*) and Grey Ducks (*A. superciliosa*). *Auk* 111, 970-978.

Rhymer JM, Williams MJ and Kingsford RT (2004) Implications of phylogeography and population genetics for

subspecies taxonomy of Grey (Pacific Black) Duck *Anas superciliosa* and its conservation in New Zealand. *Pacific Conservation Biology* 10, 57-66.

Rogers AEF (1972) NSW bird report for 1971. *Australian Birds* 6, 77-99.

Rogers AEF (1976) NSW bird report for 1976. *Australian Birds* 10, 61-84.

Schodde R (1977) Contributions to Papuasian ornithology: VI Survey of birds of southern Bougainville Island, Papua New Guinea. *CSIRO Division of Wildlife Research Technical Paper* 34, 1-103.

Short LL (1969) Taxonomic aspects of avian hybridization. *Auk* 86, 84-105.

Short LL (1978) The biological and taxonomic status of the Mexican duck. *Auk* 95, 625.

Smith J and Smith P (1990) *Fauna of the Blue Mountains.* (Kangaroo Press Pty Ltd: Kenthurst, Australia)

Thomson GM (1922) *The Naturalisation of Animals and Plants in New Zealand.* (Cambridge University Press: Cambridge)

Threatened Waterfowl Specialist Group (2003) Threatened waterfowl species and subspecies. *TWSG News* 14, 2-4.

Tracey JP, Lukins BS and Haselden C (2008) Hybridisation between Mallard (*Anas platyrhynchos*) and Grey Duck (*A. superciliosa*) on Lord Howe Island and management options. *Notornis* 55, 1-7.

Watanabe T, Mizutani M, Wakana S and Tomita T (1985) Demonstrations of maternal inheritance of avian mitochondrial-DNA in chicken-quail hybrids. *Journal of Experimental Zoology* 236, 245-247.

Whatmough RJ (1978) Birds of the Torrens River, Adelaide. *South Australian Ornithologist* 28, 1-15.

Williams M (1981) *The Duckshooter's Bag.* (The Wetland Press: Wellington, New Zealand)

Williams M and Basse B (2006) Indigenous Grey Ducks, *Anas superciliosa*, and introduced Mallards, *Anas platyrhynchos*, in New Zealand: processes and outcome of a deliberate encounter. *Acta Zoologica Sinica* 52 (Suppl.), 579-582.

Williams MJ (1969) Courtship and copulatory behaviour of the New Zealand Grey Duck. *Notornis* 16, 23-32.

Yamashina Y (1948) Notes on the Marianas Mallard. *Pacific Science* 2, 121-124.

Received 27 November 2008; accepted 9 April 2009

Mallard × Pacific Black Duck hybrids on Lord Howe Island where Pacific Black Ducks are now considered extinct. Photo by JP Tracey.

Conservation Biology: a 'crisis discipline'

Fiona Hogan and Raylene Cooke

School of Life and Environmental Sciences, Deakin University, 221 Burwood Hwy. Burwood, Victoria, Australia 3125
email: fiona.hogan@deakin.edu.au

Abstract

Conserving biodiversity is of utmost importance on a global scale. Species conservation, however, is a challenging task, which is often compounded by a lack of knowledge of target species. New advances in information technology and molecular techniques, however, are enabling conservation biologists to obtain large amounts of data quickly, which will certainly aid in assigning conservation priorities. This article reviews the use of genetics in conservation biology and highlights, using the Powerful Owl *Ninox strenua* as an example, how DNA can be a valuable source of data. (*The Victorian Naturalist* 126 (3) 2008, 92-97)

Keywords: Powerful Owl, Conservation, Surrogate Species, DNA, Feathers

Uncontrolled manipulation of the world's ecosystems has resulted in the current rate of species loss being higher than at any previous time in human history (Soule 1991). This rapid decline in species richness, and the large number of species which are facing imminent extinction (Jetz *et al.* 2007), has been the trigger for the rapidly expanding field of conservation biology, also described as a 'crisis discipline' (Soule 1985).

Crisis disciplines arise from urgency, where there is an immediate need to understand the processes causing the crisis and to obtain knowledge on how to prevent, rectify or minimise its effects (DeSalle and Amato 2004). Consequently, there is a rapid expansion of tools used to solve these problems (Meine *et al.* 2006). Conservation biology is certainly one of these disciplines. It has benefited considerably from recent advances in both information technology and molecular biology techniques, which have enabled large amounts of data to be collected, stored and analysed. A large array of software packages designed specifically to interpret molecular data have become readily available (Lowe *et al.* 2004), and novel DNA markers provide a mechanism for understanding the ecology and biology of a diverse range of wildlife (DeYoung and Honeycutt 2005).

The integration of information technology and molecular techniques has accelerated the speed and accuracy of genetic analysis (DeSalle and Amato 2004). High-through put sequencing and multi-plex genotyping allows for a large number of genetic samples to be processed simultaneously (Bertorelle *et al.* 2004). Newer geospatial technologies such as geographic information systems (GIS) are also being integrated with genetic analysis, which has given rise to yet another relatively new discipline, landscape genetics (Manel *et al.* 2003; Watts *et al.* 2004). The integration of these tools is enabling conservation biologists to collect data more rapidly, and provide improved management recommendations (DeYoung and Honeycutt 2005).

One of the major dilemmas that conservation biologists face is determining which species to conserve. It is impossible to monitor and manage every aspect of biodiversity and therefore, using a single species as a target is often adopted as a conservation tool (Simberloff 1998, Favreau *et al.* 2006). The ultimate aim of such an approach is to achieve community or ecosystem conservation by protecting a surrogate species. Surrogate species can have varying levels of ecological importance within ecosystems and therefore the identification of appropriate targets can deliver wider conservation goals (Wilcove 1993).

Keystone taxa are often selected as surrogate species (Simberloff 1998). These are species that have a critical ecological role in their ecosystem, where their disappearance has major implications beyond what might be expected, considering their biomass or abundance (Andelman and Fagan 2000). Identification of such species is beneficial for conservation, as their presence assures the ecological integrity of the communities they influence (Simberloff 1998). Keystone species, however, are difficult to identify without intricate knowledge of ecosystem dynamics, and there have been few detailed studies of keystone species (Simberloff 1998).

Surrogate species may also act as umbrellas, flagships or indicators. Umbrella species are often high-trophic-level mammalian or avian predators (Ozaki *et al.* 2006) which typically occupy large areas of habitat (Simberloff 1998). The protection of an umbrella species should theoretically save an entire suite of sympatric species with less demanding habitat requirements. Flagship species are those with high public appeal, usually large charismatic vertebrates, which are often used to promote environmental awareness (Simberloff 1998; Caro *et al.* 2004). The protection of flagship species and their habitat will lead to wider conservation benefits, where other species which share the same resources will inadvertently be protected (Andelman and Fagan 2000). Indicator species share some of the same habitat requirements as species, communities or ecosystems for which they indicate (Favreau *et al.* 2006) and therefore can be used to monitor ecosystem condition and health (Simberloff 1998). Although surrogate species are often employed by conservation biologists to help tackle conservation problems, the choice of particular surrogates is largely *ad hoc* (Landres *et al.* 1988). The use of surrogate species (umbrellas, flagships and indicators) has been found to have limited conservation benefits for protecting regional biota (Caro *et al.* 2004) and greater care in the choice of surrogate species may be required if they are to be successfully used in conservation biology (Caro and O'Doherty 1999).

Raptors as surrogate species

Humans have been fascinated with top-order predators such as raptors throughout history (Sergio *et al.* 2006). The charismatic appeal of these species has resulted in top-order predators being used as conservation targets for example, as flagship species to acquire financial support (White *et al.* 1997), raise environmental awareness (van Balen *et al.* 2000) and plan protected area systems (Andelman and Fagan 2000). Raptors are often perceived as highly sensitive with respect to their habitat and resource requirements, and therefore sensitive to habitat modification (Boal and Mannan 1999). They also have a low tolerance to disturbance (Thiollay 2006) where their breeding success, for example, is reduced by anthropogenic threats. Due to their vulnerability, raptors have been used as indicator species, where their presence can be an indica-

tion of a particular habitat quality (Sergio *et al.* 2006).

Owls are often used as flagship species for conservation campaigns, probably none more so than the Northern Spotted Owl *Strix occidentalis caurina* in the United States of America (Simberloff 1998). The mandatory requirement for the US Forest Service to use management surrogate species led to the Northern Spotted Owl becoming the flagship for the Pacific Northwest Region of the USA (Dunk *et al.* 2006). The rationale for the decision was three fold: (1) it was a threatened species; (2) it was charismatic; and (3) it was reliant on large amounts of old-growth forest (Lamberson *et al.* 1992). As the Northern Spotted Owl requires large areas of old growth forest for its survival and reproduction, it was assumed that many other species, which also rely on old growth forest, would retrospectively be protected (Lamberson *et al.* 1994). The conservation of the Northern Spotted Owl would therefore serve not only as a flagship species but also as an umbrella species.

Similarly, in Australia the Powerful Owl *Ninox strenua* has also been used as a surrogate species, particularly in regard to forestry operations and urban planning (Loyn *et al.* 2001). To help protect the Powerful Owl from the adverse impacts of timber harvesting, large amounts of forest have been reserved in Powerful Owl management areas (POMAs) to provide sufficient habitat for the owl (McCarthy *et al.* 1999). The conservation of this habitat should also protect sympatric species within these forested environments; so the Powerful Owl is serving as an umbrella species within these ecosystems.

The charismatic appeal of the Powerful Owl has led to its high public profile. It is often used as a flagship species for urban development. A recent example was in the development of a major freeway extension in Melbourne's outer eastern suburbs. The presence of the Powerful Owl in the Mullum Mullum creek corridor was used by conservationists to rally against the proposed new freeway developments. Media releases stated that ' ... the Powerful Owl lived up to its name in the eastern suburbs' because the government committed $326 million to an alternative freeway route, including a 1.5 km tunnel under the Mullum Mullum Creek corridor to protect habitat for the Powerful Owl (Tinkler 2004). While original flora and fauna

Female Powerful Owl *Ninox strenua*. Photo by Fiona Hogan

Fig. 1. The distribution of the Powerful Owl in Australia (adapted from Higgins 1999)

surveys of the Mullum Mullum creek corridor revealed that the proposed area of construction contained high intrinsic habitat value and that effects of the proposed development on flora and fauna would be considerable (Department of Conservation 1990), this report made no specific mention of the Powerful Owl. Regardless of this, the Powerful Owl was still used as a flagship species in this instance, and implementation of the report was successful in preserving valuable habitat within the urban matrix of Melbourne.

The Powerful Owl is the largest and arguably the most charismatic owl species in Australia. It is endemic to Australia and is distributed across the three eastern mainland states: Victoria, New South Wales and Queensland (Fig. 1). It has a limited distribution along the east coast (Garnett and Crowley 2000), where human population density and urban growth are also particularly high (Luck 2007).

The Powerful Owl is of international concern and is listed in Appendix II of CITES (Convention on International Trade of Endangered Species of Wild Fauna and Flora) and considered

Least Concern by the IUCN (2001 IUCN Red List of Threatened Species). Nationally, it is classified as of conservation significance (Higgins 1999) and vulnerable within the States of Victoria (Department of Sustainability and Environment 2003), New South Wales (Olsen 1998) and Queensland (Olsen 1998).

Risks to the persistence of Powerful Owls pertain largely to the loss or degradation of essential habitat (Brouwer and Garnett 1990). Traditionally perceived as a habitat and dietary specialist, the Powerful Owl was thought to require continuous tracts of old growth forest (Schodde and Mason 1980; Debus and Chafer 1994). The large body size and high metabolic rate of Powerful Owls necessitates a large, energy-rich diet, comprising medium-sized arboreal prey such as possums (Webster *et al.* 1999, Cooke *et al.* 2002a) and successful reproduction requires the presence of tree hollows suitable for nesting (Cooke *et al.* 2002b). Current research, however, suggests that the Powerful Owl is more adaptable than once perceived, inhabiting forest and woodland remnants close to major urban centres, including Brisbane, Melbourne and Sydney (Fig. 1) (Pavey 1993; Cooke *et al.* 2002a; Kavanagh 2004). It is uncertain whether owls in these urban centres are remnant populations (Kavanagh 2004) or are associated with the abundance of potential prey species and the increased protection of remnant patches within the urban matrix (Cooke *et al.* 2006).

The presence of Powerful Owls in major Australian cities is of importance, especially considering the conservation significance of this species. Surveys conducted by Miller (2003) indicated that Victorians have a relatively strong emotional attachment to individual animals and are interested in learning about wildlife and the natural environment. The presence of the Powerful Owl in urban environments, therefore, provides the opportunity to use the charismatic appeal of this species to promote environmental awareness amongst city people, and as a surrogate species for conservation. However, in order to successfully use the Powerful Owl as an umbrella, flagship or indictor species, further knowledge about its ecology, biology and habitat requirements is fundamental.

Although the Powerful Owl has been the focus of numerous studies over the past 30 years, knowledge on many aspects of its ecology and biology remain unknown. Most published studies have focused on diet (e.g. Chafer 1992; Debus and Chafer 1994; Cooke *et al.* 1997; Webster *et al.* 1999; Cooke *et al.* 2002a; Cooke *et al.* 2006) and habitat preference (e.g. Schodde and Mason 1980; Debus and Chafer 1994; Kavanagh 1998; Cooke *et al.* 2002b). Critical data on other aspects of Powerful Owl biology and ecology, such as mating systems, population structure and dispersal are currently unknown. Their dispersed distribution, low population densities, nocturnal activity cycle and difficulty in establishing the identity of individual birds has inhibited the collection of this information.

The use of genetics in conservation biology
Conservation genetics presents an opportunity to reduce current knowledge gaps in ecological studies. Genetic information can strengthen conservation knowledge and ensure that rational management decisions are made (DeSalle and Amato 2004). Microsatellites (short tandem nucleotide repeats) have become the genetic marker of choice for studies of intraspecific variation of wild populations (DeWoody

2005). A limitation of this approach is that microsatellite markers usually have to be developed for the species under investigation (Sunnucks 2000) which can be time-consuming and costly (Piggott and Taylor 2003). Once developed, however, microsatellite markers can be used in other closely related species, so that the development process does not need to be repeated for every species (Piggott and Taylor 2003).

A number of microsatellite markers used in combination, can provide a DNA profile which can unequivocally identify individuals (Piggott and Taylor 2003). Hogan *et al.* (2007) developed a suite of 14 microsatellite DNA markers from the Powerful Owl. The resolution of these markers was sufficient to provide a probability of identity (P_{ID}) of 0.0001 (1 in 10 000) for unrelated Powerful Owls, which is more than sufficient for a species with a relatively small population size (~7,000 breeding adults) and sparse distribution (Garnett and Crowley 2000). Individual DNA profiling allows for crucial elements of the breeding ecology to be assessed and the relatedness of individuals to be determined which can identify inbreeding (Galeotti *et al.* 1997). When gender is inferred by employing sex-specific markers, DNA profiling can be an extremely valuable tool, for identifying putative parents, inferring mating systems, assessing sex bias dispersal and sex-ratios of offspring.

Genetic analysis traditionally required large amounts of DNA, therefore, studies involving wild animals employed destructive sampling where the animal was killed to obtain tissue samples (Taberlet *et al.* 1999). An alternative was to capture the animal to obtain tissue samples without killing it, but this is traumatic and accidental death can occur. These two types of sampling have the advantage of providing abundant good quality DNA (Taberlet *et al.* 1999); however, neither are conducive to conservation biology. Another disadvantage of the non-destructive sampling is that capture may alter the normal behaviour of the individual being studied (Morin *et al.* 1994).

Development of the polymerase chain reaction (PCR) in the early 1980's (Saiki *et al.* 1985) enabled very small amounts of sample to be used for genetic analysis. PCR involves DNA sequence being amplified or 'copied' by enzymatic reaction *in vitro*, using short pieces of DNA (primers). Millions of copies of the target DNA sequence can be produced, which can subsequently be used for a range of genetic analyses. PCR enables the implementation of non-destructive and non-invasive sampling techniques, where DNA can be obtained from small amounts of tissue (biopsy samples), blood, feathers, hair or trace material left by the animal. PCR has therefore provided the greatest breakthrough in terms of genetics in conservation biology, as it has eliminated the need to destroy animals for research.

PCR has made possible the alternative sampling technique of non-invasive genetic sampling (NGS), where DNA left behind by an animal, such as shed hair, scats and feathers can be collected (Waits and Paetkau 2005). The attractiveness of this technique is the opportunity to obtain genetic material from free-ranging animals without having to catch, handle or even observe them (Taberlet and Luikart 1999). This technique is especially valuable when studying species that are rare, endangered or cryptic (Piggott and Taylor 2003), where using invasive study methods such as trapping is neither feasible nor appropriate (Greenwood 1996).

Shed feathers are a readily available DNA source from species which regularly moult, such as the Powerful Owl. Feathers can be collected from underneath roosts and easily identified through comparison to museum specimens. Hogan *et al.* (2008) demonstrated that a large number of DNA samples (shed feathers) can be collected over a large spatial scale, within a relatively short period of time. This mode of DNA sampling is revolutionary for ecological studies, and will provide data which otherwise would be impossible to obtain through traditional ecological techniques such as banding. The analysis of DNA extracted from samples can provide a wealth of information such as individual identification, estimates of relatedness, pedigree reconstruction, sex identification, estimates of census and effective population size, and the level of genetic polymorphism within or between populations (Taberlet and Luikart 1999; Piggott and Taylor 2003). Information obtained from such genetic data will greatly improve our knowledge about the biology and ecology of species, such as the Powerful Owl, which can further be disseminated into management strategies and subsequently enhance the effectiveness of future conservation efforts.

References

Andelman SJ and Fagan WF (2000) Umbrellas and flagships: efficient conservation surrogates or expensive mistakes? *Proceedings of the National Academy of Sciences of the United States of America* 97, 5954-5959.

Bertorelle G, Bruford M, Chemini C, Vernesi C and Hauffe HC (2004) New, flexible Bayesian approaches to revolutionize conservation genetics. *Conservation Biology* 18, 584-584.

Boal CW and Mannan RW (1999) Comparative breeding ecology of Cooper's Hawks in urban and exurban areas of southeastern Arizona. *Journal of Wildlife Management* 63, 77-84.

Brouwer J and Garnett S (1990) *Threatened birds of Australia - an annotated list.*

Caro T, Engilis A, Fitzherbert E and Gardner T (2004) Preliminary assessment of the flagship species concept at a small scale. *Animal Conservation* 7, 63-70.

Caro T and O'Doherty G (1999). On the use of surrogate species in conservation biology. *Conservation Biology* 13, 805-814.

Chafer C (1992) Observations of the Powerful Owl *Ninox strenua* in the Illawarra and Shoalhaven Regions of New South Wales. *Australian Bird Watcher* 14, 289-300.

Cooke R, Wallis R, Webster A and Wilson J (1997) Diet of a family of Powerful Owls *Ninox strenua* in Warrandyte, Victoria. *Proceedings of the Royal Society of Victoria* 109, 1-6.

Cooke R, Wallis R and Webster A (2002a) Urbanisation and the Ecology of Powerful Owls *Ninox strenua* in outer Melbourne, Victoria. In *Ecology and Conservation of Owls*, pp. 100-106. Eds I Newton, R Kavanagh, P Olsen, and I Taylor. (CSIRO: Melbourne, Australia).

Cooke R, Wallis R and White J (2002b) Use of vegetative structure by Powerful Owls in outer urban Melbourne, Victoria, Australia - Implications for management. *Journal of Raptor Research* 36, 294-299.

Cooke R, Wallis R, Hogan F, White J and Webster A (2006). The diet of Powerful Owls *Ninox strenua* and prey availability in a continuum of habitats from disturbed urban fringe to protected forest environments in south-eastern Australia. *Wildlife Research* 33, 199-206.

Debus S and Chafer C (1994) The Powerful Owl *Ninox strenua* in New South Wales. *Australian Birds* 28, S21-S64.

Department of Conservation E. a. L (1990) Flora and fauna of the Koonung and Mullum Mullum Valleys (Proposed Eastern arterial road and Ringwood bypass) Victoria. Department of Conservation, Forests and Lands, Melbourne, Australia.

Department of Sustainability and Environment (2003) Advisory list of threatened vertebrate fauna in Victoria. Department of Sustainability and Environment, Melbourne.

DeSalle R and Amato G (2004) The expansion of conservation genetics. *Nature Reviews Genetics* 5, 702-712.

DeWoody JA (2005) Molecular approaches to the study of parentage, relatedness, and fitness: Practical applications for wild animals. *Journal of Wildlife Management* 69, 1400-1418.

DeYoung RW and Honeycutt RL (2005) The molecular toolbox: Genetic techniques in wildlife ecology and management. *Journal of Wildlife Management* 69, 1362-1384.

Dunk JR, Zielinski WJ and Welsh HH (2006) Evaluating reserves for species richness and representation in northern California. *Diversity and Distributions* 12, 434-442.

Favreau JM, Drew CA, Hess GR, Rubino MJ, Koch FH and Eschelbach KA (2006) Recommendations for assessing the effectiveness of surrogate species approaches. *Biodiversity and Conservation* 15, 3949-3969.

Galeotti P, Pilastro A, Tavecchia G, Bonetti A and Congiu L (1997) Genetic similarity in Long-eared Owl communal winter roosts: a DNA fingerprinting study. *Molecular Ecology* 6, 429-435.

Garnett ST and Crowley GM (2000) *The Action Plan for Australian Birds*. (Environment Australia: Canberra)

Greenwood J (1996) Basic techniques. In *Ecological Census Techniques, A Handbook* pp. 11-110. Ed W. Sutherland. (Cambridge University Press: Cambridge).

Higgins PJE (1999) *Handbook of Australian, New Zealand and Antarctic Birds. Volume 4: Parrots and Dollarbird*. (Oxford University Press: Melbourne)

Hogan F, Burridge C, Cooke R and Norman J (2007) Isolation and characterisation of microsatellite loci to DNA fingerprint the Powerful Owl *Ninox strenua*. *Molecular Ecology Notes* 7, 1305 - 1307.

Hogan F, Cooke R, Burridge C and Norman J (2008) Optimizing the use of shed feathers for genetic analysis. *Molecular Ecology Resources* 8, 561-567.

Jetz W, Wilcove DS and Dobson AP (2007) Projected impacts of climate and land-use change on the global diversity of birds. *Plos Biology* 5, 1211-1219.

Kavanagh RP (1998) Ecology and management of large forest owls in southeastern Australia. *Australian Journal of Ecology* 23, 184-185.

Kavanagh RP (2004) Conserving owls in Sydney's urban bushland: current status and requirements. *Urban Wildlife* 93-108.

Lamberson RH, McKelvey R, Noon BR and Voss C (1992) A dynamic analysis of Northern Spotted Owl viability in a fragmented forest landscape. *Conservation Biology* 6, 505-512.

Lamberson RH, Noon BR, Voss C and McKelvey KS (1994) Reserve design for territorial species - the effects of patch size and spacing on the viability of the Northern Spotted Owl. *Conservation Biology* 8, 185-195.

Landres PB, Verner J and Thomas JW (1988) Ecological uses of vertebrate indicator species - a critique. *Conservation Biology* 2, 316-328.

Lowe A, Harris S and Ashton P (2004) *Ecological Genetics: Design, Analysis and Application*. (Blackwell Publishing, Victoria, Australia).

Loyn RH, McNabb EG, Volodina L and Willig R (2001) Modelling landscape distributions of large forest owls as applied to managing forests in north-east Victoria, Australia. *Conservation Biology* 97, 361-376.

Lack GW (2007) The relationships between net primary productivity, human population density and species conservation. *Journal of Biogeography* 34, 201-212.

Manel S, Schwartz MK, Luikart G and Taberlet P (2003) Landscape genetics: combining landscape ecology and population genetics. *Trends in Ecology and Evolution* 18, 189-197.

McCarthy M A, Webster A, Loyn RH and Lowe KW (1999) Uncertainty in assessing the viability of the Powerful Owl *Ninox strenua* in Victoria, Australia. *Pacific Conservation Biology* 5, 144-154.

Meine C, Soule M and Noss RF (2006) "A mission-driven discipline": the growth of conservation biology. *Conservation Biology* 20, 631-651.

Miller KK (2003) Public and stakeholder values of wildlife in Victoria, Australia. *Wildlife Research* 30, 465-476.

Morin PA, Wallis J, Moore JJ and Woodruff DS (1994) Paternity exclusion in a community of wild chimpanzees using hypervariable simple sequence repeats. *Molecular Ecology* 3, 469-477.

Olsen P (1998) *Australia's Raptors: Diurnal Birds of Prey*. (Birds Australia: Melbourne).

Ozaki K, Isono M, Kawahara T, Iida S, Kudo T and Fukuyama K (2006) A mechanistic approach to evaluation of umbrella species as conservation surrogates. *Conservation Biology* 20, 1507-1515.

Pavey CR (1993) The Distribution and Conservation Status of the Powerful Owl (*Ninox strenua*) in Queensland. In *Australian Raptor Studies*, pp. 144-153. Ed P Olsen (Australian Raptor Association Melbourne: Australia).

Piggott MP and Taylor AC (2003) Remote collection of animal DNA and its applications in conservation management and understanding the population biology of rare and cryptic species. *Wildlife Research* 30, 1-13.

Saiki RK, Scharf S, Faloona F, Mullis KB, Horn GT, Erlich HA and Arnheim N (1985) Enzymatic amplification of beta-globin genomic sequences and restriction site analysis for diagnosis of sickle-cell anemia. *Science* 230, 1350-1354.

Schodde R and Mason IJ (1980) *Nocturnal Birds of Australia* (Lansdowne Editions: Melbourne)

Sergio F, Newton I, Marchesi L and Pedrini P (2006). Ecologically justified charisma: preservation of top predators delivers biodiversity conservation. *Journal of Applied Ecology* 43, 1049-1055.

Simberloff D (1998) Flagships, umbrellas, and keystones: is single-species management passe in the landscape era? *Biological Conservation* 83, 247-257.

Soule ME (1985) What is conservation biology? *Bioscience* 35, 727-734.

Soule ME (1991) Conservation - tactics for a constant crisis. *Science* 253, 744-750.

Sunnucks P (2000) Efficient genetic markers for population biology. *Trends in Ecology and Evolution* 15, 199-203.

Taberlet P and Luikart G (1999) Non-invasive genetic sampling and individual identification. *Biological Journal of the Linnean Society* 68, 41-55.

Taberlet P, Waits LP and Luikart G (1999) Noninvasive genetic sampling: look before you leap. *Trends in Ecology and Evolution* 14, 323-327.

Thiollay JM (2006) The decline of raptors in West Africa: long-term assessment and the role of protected areas. *Ibis* 148, 240-254.

Tinkler C (2004) $400m bill for creature comfort. *Sunday Herald Sun*, Melbourne. 29

van Balen S, Nijman V and Prins HHT (2000) The Javan hawk-eagle: misconceptions about rareness and threat. *Biological Conservation* 96, 297-304.

Waits LP and Paetkau D (2005) Noninvasive genetic sampling tools for wildlife biologists: a review of applications and recommendations for accurate data collection. *Journal of Wildlife Management* 69, 1419-1433.

Watts P C, Rouquette JR, Saccheri J, Kemp SJ and Thompson DJ (2004) Molecular and ecological evidence for small-scale isolation by distance in an endangered damselfly, (*Coenagrion mercuriale*). *Molecular Ecology* 13, 2931-2945.

Webster A, Cooke R, Jameson G and Wallis R (1999) Diet, roosts and breeding of Powerful Owls *Ninox strenua* in a disturbed, urban environment: a case for cannibalism? Or a case of infanticide? *Emu* 99, 80-83.

White PCL, Gregory KW, Lindley PJ and Richards G (1997) Economic values of threatened mammals in Britain: a case study of the Otter *Lutra lutra* and the Water Vole *Arvicola terrestris*. *Biological Conservation* 82, 345-354.

Wilcove D (1993) Getting ahead of the extinction curve. *Ecological Applications* 3, 218-220.

Received 3 November 2008; accepted 17 May 2009

BOCA Western Port Survey: long-term monitoring of waterbird numbers

Xenia Dennett[1] and Richard H Loyn[2]

[1] Bird Observation & Conservation Australia, PO Box 185, Nunawading, Victoria 3131
[2] Arthur Rylah Institute, Department of Sustainability and Environment, 123 Brown St, Heidelberg, Victoria 3084

Abstract

Since 1973 volunteers from the Bird Observers Club of Australia (BOCA, formerly known as Bird Observers Club), have counted birds several times a year at their high-tide roosts within Western Port, a Ramsar designated site. The survey focuses on waterbirds and shorebirds in the tidally influenced areas. This paper presents data for some of the species that have declined, increased or remained constant in numbers during this period and mentions some factors responsible for these numbers. (*The Victorian Naturalist* 126 (3). 2009, 99-107)

Keywords: waterbirds; shorebirds; survey; Western Port; Bird Observers Club of Australia

Introduction

Western Port is a large shallow embayment east of Melbourne in south-east Australia. It is one of the three most important sites for shorebirds in Victoria (Watkins 1993) and is a Ramsar site. Most of the bay is shallow, has extensive mudflats at low tide and is fringed by mangroves (*Avicennia marina*) and saltmarsh. There is some industry and commercial shipping centred around Hastings in the west with a deepwater port and shipping channels maintained by dredging.

Western Port has three marine parks and is very popular for recreational boating and fishing. Phillip Island, at the southern edge of the Bay, is a major tourist attraction, particularly in summer. French Island in the centre of the Bay is twice as large as Phillip Island, but it is sparsely populated: the French Island National Park covers two-thirds of the island. Both islands have extensive freshwater and brackish wetlands.

Western Port provides valuable roosting and foraging habitats for shorebirds and waterbirds with rich areas of mangroves, saltmarsh and intertidal mudflats. Western Port has four wetlands types: permanent saline and semi-permanent saline around the Bay, deep freshwater marshes and permanent open freshwater areas on French and Phillip Islands.

The BOCA Western Port Survey, initiated by Richard Loyn in 1973, began with the aim of locating major high-tide roosts and counting birds whilst they were concentrated at those roosts. At low-tide, waterbirds are widely dispersed over vast areas of mudflats, making regular counts impractical. The survey continues

to focus on waterbirds and shorebirds in the tidally influenced areas of the Bay (Loyn 1978; Dann *et al*. 1994; Loyn *et al*. 1994; Loyn 2002; Loyn and Dennett 2008).

Since 1973, teams of BOCA volunteers and others, have gathered on a predetermined day at some 20 regular sites within the Bay, plus other conditional sites, prior to the daytime high tide. All waterbirds are counted and numbers recorded. Counts are made at least three times a year, in February when the Palaearctic shorebirds are preparing to migrate to their breeding grounds in the far Northern Hemisphere and when others use Western Port as a summer refuge; in June/July for resident and over-wintering migrants and in November/December when post breeding and juvenile Palaearctic shorebirds have returned to moult and spend the Austral summer.

As a consequence of this continuous, long-running survey the significance of Western Port for birds has become widely recognised. It is listed under the Ramsar Convention on Wetlands and is part of the Shorebird Site Network for the East Asian Australasian Flyway. In 2002, it was included as part of the Biosphere Reserve for Western Port and Mornington Peninsula, has been recognized by WWF for Nature Australia Shorebird Conservation Project and is a key monitoring site for Palaearctic and Australasian shorebirds.

The wetland avifauna of Western Port can be broadly categorised as:
• waterbirds, including the fishers (cormorants, pelican and terns), waterfowl (ducks, grebes, swan, etc.), crakes and rails, gulls and large

wading birds (herons, ibis, spoonbills and egrets);
- international migratory shorebirds, comprising Palaearctic species breeding in Asia and the Arctic and one Australasian species, Double-banded Plover, breeding in mountains of the South Island of New Zealand;
- Australian breeding shorebirds, including dotterels, plovers, avocet, stilts, oystercatchers and lapwings; and
- bushbirds, associated with wetlands, such as songbirds, raptors, kingfishers and parrots.

To satisfy the Ramsar Convention criteria for identifying areas of international importance for shorebirds in Australia, an area must support one per cent or more of the world population of a particular species or subspecies of shorebirds, or support 20 000 or more individuals of that species or subspecies. As a result of the BOCA survey and some additional studies Western Port was recognised as having internationally significant numbers of Eastern Curlew, Common Greenshank, Curlew Sandpiper, Red-necked Stint and Double-banded Plover and a significant national population of Pacific Golden Plover (Watkins 1993).

The smallest and most numerous Palaearctic waders in Western Port and Australia are Red-necked Stints and Curlew Sandpipers. Observations of these small waders, banded by Victorian Wader Study Group, show their migration routes from Western Port to Siberian breeding grounds occur in four or five stages, with 'refueling' stopovers in north-west Australia (Broome), south-east Asia, further north in eastern China and perhaps inland Siberia. Larger waders, Bar-tailed Godwits and Red Knot make fewer stopovers (Minton *et al.* 2006).

Overseas factors such as loss of feeding sites due to land reclamation (as at Saemangeum in Korea), reduction in water flows due to dam formations, and weather events causing early or late thaws/melts, may all impact on the environment of migrating birds.

What other factors may be occurring? One local event in the early 1980s, occurring throughout Western Port for unknown reasons, was a marked decline in sea grass (*Zostera* spp). Sea grass is essential as a primary food source for some species, such as the herbivorous Black Swan, and provides essential protective nurs-

ery cover for fingerlings and other aquatic species. Hence 'fishers' such as Pied and Little Pied Cormorant and Australian Pelican all showed similar significant decreases following this very serious local event.

Throughout Western Port the sea beds are variable, with significant regional differences of sand, silt and mud substrates and water flows. Various species take advantage of these different niches. Long term monitoring over three seasons each year has shown interesting trends in shorebird numbers. Some examples follow.

Seasonal and cyclic changes
All species show some seasonal changes on an annual basis. The most marked are obviously the migrating shorebirds, but most other species also show consistent seasonal variation due to climatic and breeding requirements. For example, Black Swans are present throughout the year, but their numbers fluctuate and show a seasonal cycle. Their counts are lowest during winter and early spring (Figs. 1 and 2) when they are breeding locally on adjacent farmlands and swamps. Most locally breeding waterbirds show a seasonal pattern with minimal numbers in the winter-spring breeding season and maximum numbers in summer or autumn. Exceptions include the Great and Little Black Cormorants, which tend to be recorded most in spring.

Many of the international waders breed in Siberian regions where lemmings and voles also breed. These small rodents show cyclical 'boom and bust' breeding events. When their numbers are high, predators such as owls and foxes, have plenty of food for their young, but when the lemming numbers crash they seek other food sources, including eggs and chicks of waders. Consequently there is a reduction in juvenile waders migrating southwards. These events tend to have a periodicity of about three years. These trends and others (Loyn and Dennett 2008, Fig. 3) can be seen in some of the Western Port data, for small waders such as Curlew Sandpipers and Red-necked Stints.

Early decreases
Grey-tailed Tattler
This Palaearctic wader is most commonly found overwintering in northern Australia, with Victoria and Tasmania at the end of its range. Since the survey began there has been a dramatic decrease in their numbers. In the 1970s, 60-81

Fig. 1. Black Swans breed in local fresh water swamps and dams. Photo by Xenia Dennett.

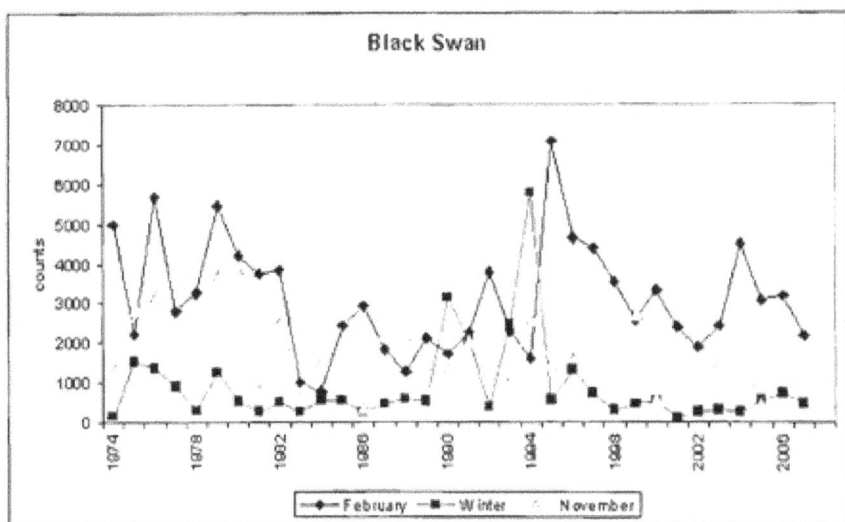

Fig. 2. Black Swan counts from BOCA Western Port Survey from 1974 to 2007 showing seasonal variations with maximum numbers in February, the summer refuge season.

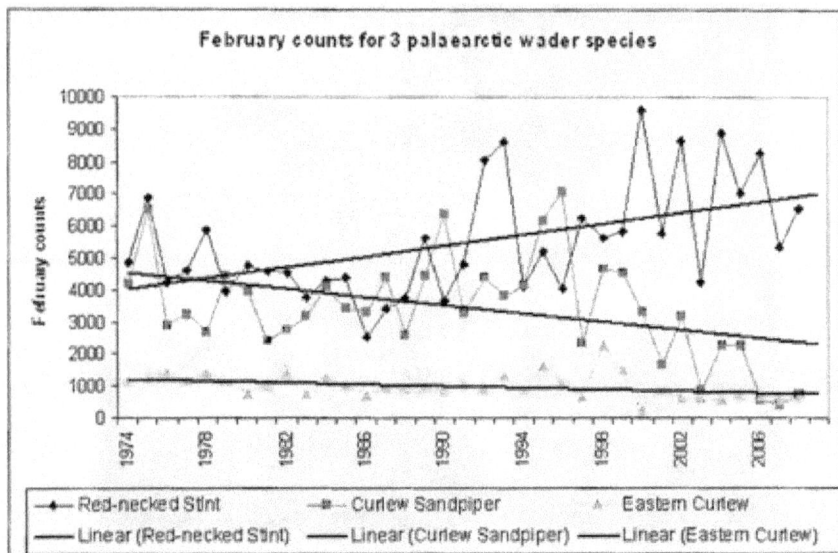

Fig.3. Total February counts for three species of palaearctic waders during the BOCA Western Port Survey period 1974 -2007. Trendlines have been added to show an apparent increase in Red-necked Stint, a significant decrease of the slightly larger Curlew Sandpiper, and a slow steady decline in the numbers of the largest wader Eastern Curlew.

individuals were counted regularly in Western Port on their return from the Arctic. In the 1980s numbers dropped to around 10, by 1990s only two and in the last decade there have been very few records. However, there has also been a similar decrease throughout northern Australia (Kearney *et al.* 2008) and the East-Asian Australasian Flyway, so local events in Western Port are unlikely to be a major factor in the local decline. These birds often roost in mangroves (making counting difficult) as well as intermittently moving their roosting sites (Minton 2000). It is therefore possible that they, and other birds, may sometimes be overlooked.

Silver Gulls

These scavenging omnivores are present throughout the year, with highest numbers in November and February. In the 1970s their numbers were increasing, until the open rubbish tip at Hastings was closed. A significant decrease followed, and as other regional tips were also closed their numbers have declined (Fig. 4). Active management measures have also delayed their recovery.

Black Swans feed directly on the sea grass growing in shallow waters, and the sudden loss of sea grass, in the early 1980s, severely affected numbers. Since then, there has been some increase in sea grass and local numbers of Black Swan are again increasing around Rhyll and Observation Point. But some beds in Western Port still show no signs of recovery, for example around the north-east section adjacent to Yallock Creek (BOCA 2003), where water turbidity in this area is very high. Is this reducing the amount of sunlight essential for photosynthesis and plant growth?

Australian Pelican numbers in February also dropped from an average of 200 prior to the sea grass loss in 1983, to about 50 a decade later, and are still continuing to struggle upwards (Fig. 5). These 'fishers' breed in inland Australia in saline waterways, and there have been significant breeding events since then at several major inland water bodies. Could these breeding events have 'depleted' the coastal numbers?

Numbers of Little Pied Cormorant, another fisher affected by the sea grass crash, showed similar declines and have not recovered (Figs.

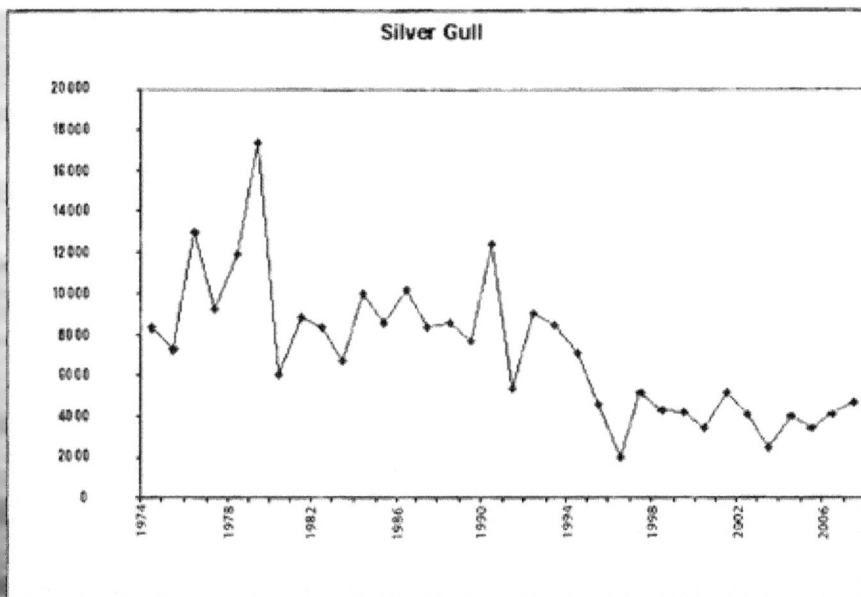

Silver Gull

Fig. 4. Silver Gull combined counts for February, winter and November, show increasing numbers until local open tip at Hastings was closed in late 1970s, followed by further regional closures in 1994.

6 and 7). The larger Pied Cormorant declined initially, but showed a later recovery. These birds are able to fish in deeper waters outside Western Port.

Recent decreases
At least two species of international migratory wader have decreased in recent years, as indicated below. These declines are usually considered to be related to events on the breeding grounds or events such as loss of habitat on their east Asian migration routes.

Curlew Sandpiper
Of all the migrating wader species these small waders (Fig. 8) have shown the largest decline throughout the East Asian-Australasian Flyway. In recent years a lack of juvenile birds in recapture studies is indicative of a significant decline in breeding success, in addition to the normal cyclical pattern (Fig. 3). The reasons for this are not known.

Eastern Curlew
Since the 1980s all counts in south-east Australia have shown a continuing decline (Gosbell and Clemens, 2006). Numbers in Western Port

seemed to hold up well until the early 1990s (Dann *et al.* 1994), but have declined subsequently (Fig.3). Probably a variety of factors are responsible, including events en route and on the breeding grounds.

General increases
Red-necked Stint
Observations on this, the smallest migrating wader, do not show the obvious downward trend shown by most other Palaearctic waders, and there even appears to be an increase (Fig. 3) in numbers. This is in striking contrast to the slightly larger, but still very small, Curlew Sandpiper mentioned above. Both have a rather similar northern migration route, although Curlew Sandpipers tend to take a more western route on their southward migration (Minton *et al.* 2006). Why are these trends occurring?

Bar-tailed Godwit
Although most data from the east coast show a long-term decline in numbers, Western Port figures are more encouraging with a general upward trend to approximately double the numbers found in early years (Fig. 9). Why is this?

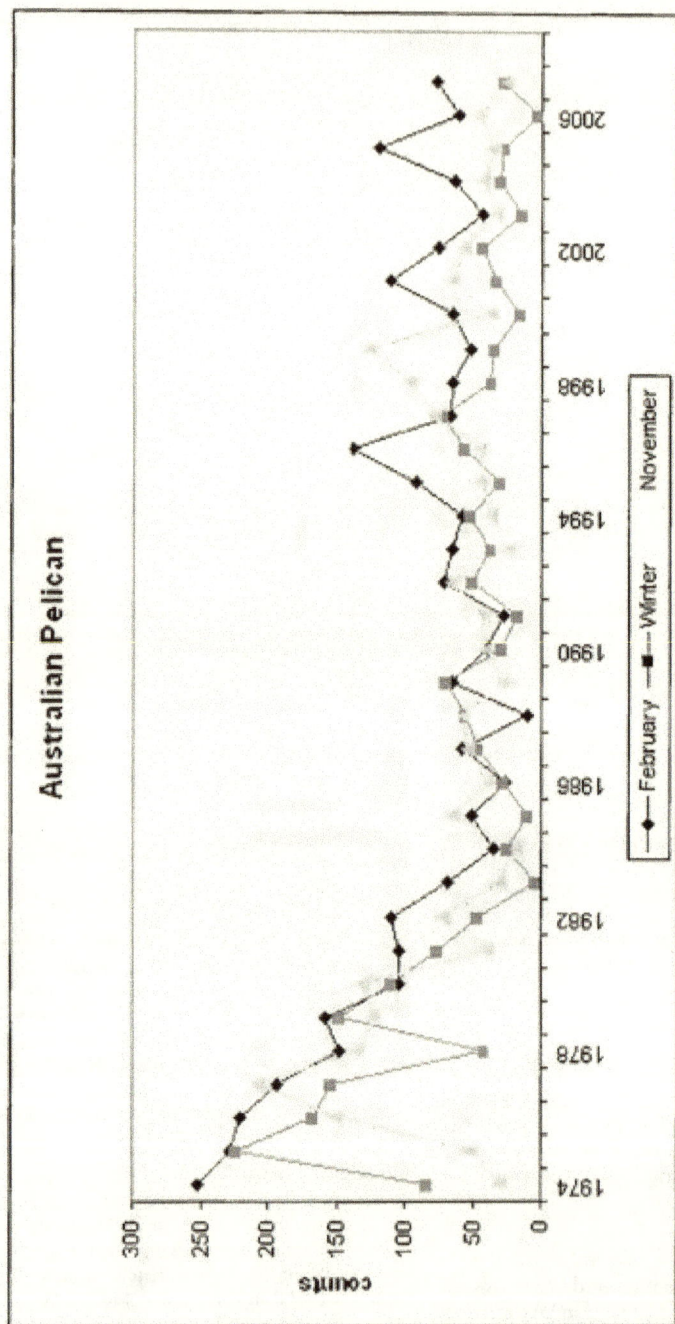

Fig. 5. Australian Pelican showing seasonal variations. Numbers were declining prior to the sea grass collapse in early 1980s and still have not recovered.

Fig. 6. Little Pied Cormorant feeds on fish and is heavily dependent on food stocks in shallow regions. Photograph from BOCA collection.

Little Pied Cormorant

February — winter — November

Fig. 7. Little Pied Cormorant counts show some seasonal variation each year, but numbers decreased after sea grass losses in 1980s, with little recovery.

Recent arrivals

Red-necked Avocets have made a dramatic appearance in Western Port counts in recent years. These are Australian endemic, breeding waders, with a characteristic upturned bill used to sieve their aquatic food of small crustaceans, worms and molluscs. They rarely oc- curred in the bay until June 1991, when seven Red-necked Avocets were counted in the Pioneer Bay area. No further birds were recorded until July 1994 when 760 birds were counted in the same area, and at the September count they numbered 850 before falling to 370 in November 1994. No birds were present in the following

Fig.8. Some Curlew Sandpipers migrate from Siberian breeding grounds to spend the austral summer in Western Port where they moult their worn plumage and 'fatten-up' before returning to breed in the following year. Photo by Xenia Dennett.

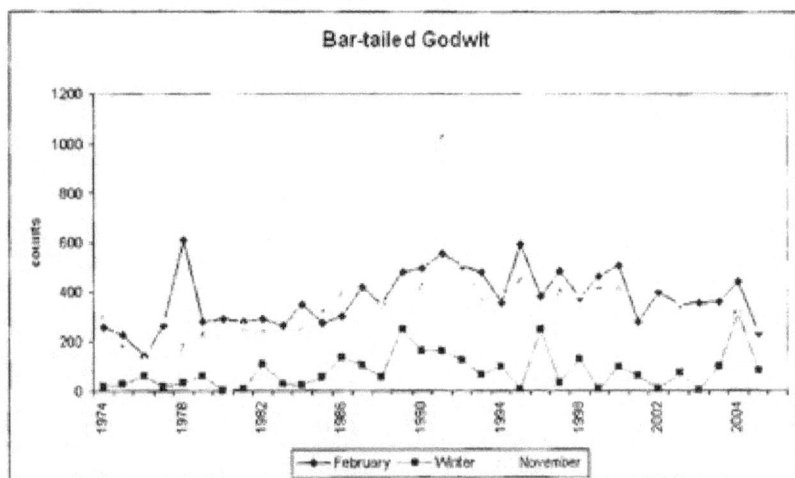

Fig. 9. Bar-tailed Godwit is one of the few species of palaearctic wader to show an increase in numbers in Western Port.

February. Since then this pattern of birds occurring in Pioneer Bay from winter through to November has become established, except in 2000 and 2007 when no avocets were counted. Because dates for the counts vary somewhat, depending on suitable tides and days, we may miss early or later arrivals as was possibly the case in June 2000 (count done 17 June). We can speculate on other possible reasons, but favourable food sources at appropriate water levels and a protected habitat are probably very important.

Other data from long-term avifauna studies have recently been published (Olsen 2008), much of which, as is the case with our data, has been obtained by dedicated volunteers. The BOCA Western Port Survey has already provided much useful information, particularly relating to seasonal and longitudinal trends. Some factors are known to be operating at the local, regional, national and international levels whilst others are unknown. But without these data one cannot begin to ask questions, plan, and attempt to manage this very complex, fragile and fascinating environment for all who use it. People, pollution and pests probably pose the greatest risks to the biodiversity of Western Port and its future, including those so very dependent upon it for food and shelter—the avifauna.

Acknowledgements

It is a pleasure to thank all the many volunteer observers who have taken part over so many years, and especially to the BOCA coordinators and other supporters, friends and colleagues and also to Laurie Living for maintaining the database.

References

BOCA (2003) *Wings over Western Port: three decades surveying wetland birds 1973-2003*. (Bird Observers Club of Australia, Report no.10).

Dann P, Loyn RH and Bingham P (1994) Ten years of water bird counts in Western Port, Victoria, 1973-83. II. Waders, gulls and terns. *Australian Bird Watcher* 15, 351-363.

Gosbell K and Clemens R (2006) Population monitoring in Australia: some insights after 25 years and future directions. *The Stilt* 50, 162-175.

Kearney B, Haslem A and Clemens R (2008) Report on population monitoring counts 2007 and summer 2008. *The Stilt* 54, 54-68.

Loyn RH (1978) A survey of birds in Westernport Bay, Victoria, 1973-74. *Emu* 78, 11-19.

Loyn RH. (2002) Changes in numbers of water birds in Western Port, Victoria, over quarter of a century (1973-1998). In *Le Naturaliste in Western Port 1802-2002*. Eds N and P Macwhirter, Sagliocco JL and Southwood J Commemorative Seminar, 13-14 April 2002.

Loyn RH, Dann P and Bingham P (1994) Ten years of waterbird counts in Western Port, Victoria, 1973-83; I. Waterfowl and large wading birds. *Australian Bird Watcher* 15, 333-350.

Loyn RH and Dennett X. Waterbirds of Western Port. (2008) In *The State of Australia's Birds 2008*, pp 25-26. Ed Penny Olsen (2008) Supplement to Wingspan, 18, no. 4, December.

Minton C (2000) Waders roosting in mangroves. *The Stilt* 37, 23-24.

Minton C, Wahl J, Jessop R, Hassell C, Collins P and Gibbs H. (2006) Migration routes of waders which spend the non-breeding season in Australia. *The Stilt* 50, 135-157.

Olsen P (2008) *The State of Australia's Birds 2008*. Supplement to Wingspan, 18, no. 4, December.

Watkins D (1993) *A National Plan for Shorebird Conservation in Australia*. Australasian Shorebird Studies Group, Royal Australasian Ornithological Union and World Fund For Nature. RAOU Report No. 90.

Received 15 January 2009; accepted 4 June 2009

One Hundred Years Ago

BIRD DAY—The first Bird Day instituted by the Education Department for the benefit of the scholars attending the State schools of Victoria was celebrated on Friday, 29th October. A special programme of lessons, dealing with various aspects of bird life and laying particular stress on their protection and preservation, was carried out. In the smaller schools, where possible, excursions were made to localities frequented by birds, and lessons given in the field. In the metropolis, as this was impossible, the services of a number of natural history enthusiasts were secured, and practical demonstrations on the value of birds were given by members of the Field Naturalists' Club, the Ornithologists' Union, and the Bird Observers' Club, to the senior classes of the suburban schools. On the whole, the movement was a success, and it is hoped will lead to more interest in our feathered friends, without whose aid human life would almost become impossible. Advantage was taken of the day to inaugurate the Gould League of Bird Lovers, and some 50,000 children handed in their names as willing to observe its precepts. We trust this movement will, as the years roll by, greatly lessen the destructive pot-hunting which takes place on every holiday.

From *The Victorian Naturalist* XXVI, p. 95, November 1909

Leigh Desmond Ahern

14 December 1951 – 7 February 2009

Leigh Ahern and his wife Charmian died defending their Steels Creek home in the Victorian bushfires of 7 February 2009. Leigh's life was a testament of dedicated, thoughtful, ecological and conservation work, based in, but not bounded by, science as he sought effective ways to nurture community involvement in conservation. Leigh was born in Ringwood and attended local Government primary and secondary schools. He went on to study at La Trobe University, where he majored in Zoology, and graduated with BSc (Hons) in 1974. Leigh's honours research project was on the White-footed Dunnart *Sminthopsis leucopus* at Sandy Point, Western Port. It remains a seminal study of this little-known mammal. Leigh began work at the then National Museum of Victoria (NMV) in March 1975. He married Charmian Nathanielsz in January 1977 and together they enjoyed a loving relationship centred on their children, pets and home in the bush adjacent to Kinglake National Park.

At the NMV, Leigh worked in the Biological Survey Department where he studied the limnology of Gippsland rivers and how it might be affected by proposed dams. He also worked with Alan Yen in pioneering studies of the invertebrate fauna of leaf litter in eucalypt and pine forests. Some of us first really got to know Leigh when he was recruited to the Fisheries and Wildlife Department's Wildlife Survey Team at Arthur Rylah Institute in Heidelberg —on 6 February 1978. The Survey team spent several years in the Gippsland bush, working in rotating pairs, systematically surveying the flora and fauna as part of the Land Conservation Council of Victoria's reviews of the use of Crown Land across the State. Out in the bush on two-week field trips we all developed a comradeship with Leigh.[1] In this 'bush apprenticeship' lugging Elliot and cage traps, spotlighting deep into the night, up early to clear the traps then conducting bird and reptile surveys through all weather—Leigh's breadth of knowledge, thoroughness and his basic compassion for both the bush and humanity inspired all of

us. Any field data marked 'LDA' was the gold standard of those times.

Leigh and Chris Belcher were quickly dubbed 'the Robur twins', in reference to their fondness for a cuppa brewed in a billy over the campfire at any opportunity. Leigh was also an accomplished musician and singer and beguiled the team on more than one occasion with folk songs. Over subsequent years, phone conversations might be peppered with reminiscences like ' … remember the Moroka trip … don't go back … it's been clear-felled'. Leigh's capacity to remember in detail events from those far-off wildlife survey days was phenomenal and this capacity for recall must have added greatly to the value of his subsequent environmental writings and advocacy.

Back in the office, pinned near Leigh's phone, was an illustration of the differences in canine length of the vespertilionid bats. Leigh had carefully annotated the Victorian Public Service classifications under each species—the longer the fangs the higher the classification! That was Leigh at his driest: humour befitting the cartoonist Leunig, and which was open to everyone's view, be it the Minister or work colleagues.

Leigh moved from research into Wildlife Management where he was the key driver in establishing the first authoritative list of Victorian threatened species; he actively contributed to many threatened species projects on birds, mammals and invertebrates. Leigh showed his breadth of knowledge of fauna when he took on the position of convenor of the Eltham Copper Butterfly Working Group.

Again ahead of his time, Leigh worked closely with the Bird Observers Club in devising and establishing the *Land for Wildlife* program *within* the Victorian Government, an innovation that has significantly improved off-reserve conservation and habitat management throughout Victoria. It has since been adopted by several other States, but alas, is now sadly neglected in its home State, for reasons that are unclear .

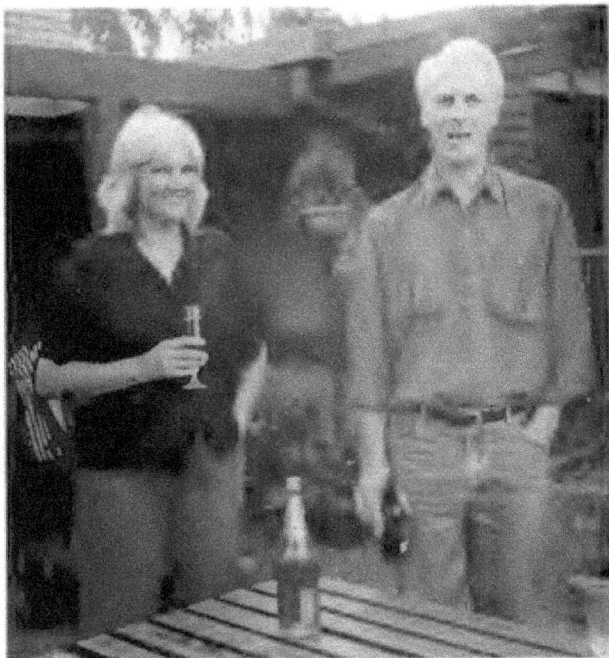

Leigh and Charmian Ahern

In December 1995 when threatened with relocation to the city, Leigh left the Department to establish a consultancy, Nature Scope, and to work from home in his beloved Kinglake bush. Here he continued his work at conserving Victorian landscapes, leading and authoring the cutting-edge bio-regional conservation plans. Leigh also took on other work to keep in contact with fellow scientists. He enjoyed a short period of work at the Museum in the late 1990s when he was employed to assist the late John Coventry to pack the herpetology collection for its move from the Abbotsford annexe to the new museum site in the city. Leigh was also employed by the Museum to prepare a list of Melbourne wildlife that formed the basis of the Museum book, *Melbourne's Wildlife*. He also undertook field work for the Museum during this period, including a trip with John Coventry to Queensland, and a trip to Uluru with John Wainer. In 2005, Leigh was employed by the Department of Primary Industries in a major project to translocate a population of Giant Gippsland Earthworms (which he believed should be named the 'Gippsland Giant Earthworm' to be grammatically correct) that

was threatened by realignment of the South Gippsland Highway. This translocation work stimulated his interest in the giant earthworm and he eagerly accepted the role of field assistant on many 'Gippsland Giant Earthworm' projects over the next three years. Leigh was also involved on field work associated with the Parks Victoria box-ironbark thinning experiment near Bendigo.

To everyone who worked with him, Leigh was a practical, thoughtful field worker with a keen, dry sense of humour. When undertaking the laborious tasks of field work—digging holes searching for giant earthworms, washing Elliot traps or cold and fruitless spotlighting—one could not want for better company. His knowledge of the land was amazing and he had an uncanny ability to engage landowners and make them feel at ease. He 'connected' and was always fascinated in those little things around him; his astute observational skills contributed to the breadth of his knowledge and compassion. An unidentified contraption of steel left abandoned in a paddock would hold his attention for ages as he studied it in an attempt to understand its use and its history. He would not let go until he was satisfied with its origins, even if it meant researching it once he got home.

Leigh's breadth of knowledge extended well outside the area of natural history, and he joined the Yarra Glen and District Historical Society in 2000. He became an active member and applied his editing skills to the local history records of other members. Between 2003 and 2007, Leigh edited eight small books on recollections of life in Steels Creek, Tarrawarra, Yarra Glen, the Maroondah aqueduct, and the Healesville-Yarra Glen rail service and stations. These books were also published as CDs. He was working on other historical books at the time of his death.

Over this time, Leigh published more than 30 scientific, policy and popular reports and papers in the literature, including a number in *The Victorian Naturalist* (see bibliography). He was a member and supporter of the Field Naturalists Club of Victoria of many years' standing.

In October 2007 the Government appointed Leigh as a member of the Scientific Advisory Committee under the *Flora and Fauna Guarantee Act*, where his wisdom and expertise on Victorian vertebrate fauna and terrestrial communities of flora or fauna and potentially threatening processes added to the committee's breadth and depth. More recently, Leigh had been invited to be a member of the Australian Natural History Medallion Award Committee. Unfortunately the conflagration of 7 February 2009 cut tragically short the continuance of Leigh's contribution to the Victorian community.

Leigh dedicated his life and career to protecting the natural beauty of Victoria and showing us how to 'live' with it and within it. It is a tragic irony that fire—one of the drivers of our landscape, yet potentially one of the 'terrors' of this wide brown land—took his life, and that of his beloved Charmian. They will both be sorely missed by all who knew them. Leigh and Charmian are survived by Dale and Chloe who made the most excellent choice of parents, and by grand-daughter Charlotte.

ᴵ In the Fisheries and Wildlife Department Research Branch of the times, budgetary constraints meant a choice between a winch or radio—a winch on 'the Tojo' made us self reliant, not needing to 'bother' local Forests Commission officers with radio contact. (The 1970s were a long way from

current OHS requirements.) The camaraderie of those times was reinforced as Leigh knew we were bringing a new ethos, based on ecology, to the forested landscapes of Gippsland for Victorians to re-evaluate land-use and management. Over many two-week field trips, 10-14 hour days were spent trapping, spotlighting, observing and recording—no overtime requested or offered, but the campfires, the occasional surf, the freedom, excitement and importance of the 'work' infected Leigh and us all.

Bibliography of publications and unpublished reports by Leigh Ahern

Ahern LD (1974) A trapping study to determine the habitat utilisation and biology of small mammals at Sandy Point, Westernport, with particular emphasis upon *Sminthopsis leucopus* (Gray). Unpublished B.Sc (Hons.) Thesis: LaTrobe University, Melbourne.

Ahern LD (1982) Threatened wildlife in Victoria and issues related to its conservation. *Fisheries and Wildlife Paper 27* Fisheries and Wildlife Division, Victoria.

Ahern LD (1983) White-footed Dunnart (*Sminthopsis leucopus*). In *Complete Book of Australian Mammals*, p. 53 Ed RG Strahan (Angus and Robertson: Sydney)

Ahern LD (1985) Clearing in the Mallee Study Area—a perspective on Mallee fauna. *Parkwatch* 142, 2-6.

Ahern LD (1998) Insect farming in Queensland—a model for Land for Wildlife. *Land for Wildlife News* 3: 12–13.

Ahern L (2001) A 'bite' from the past. *The Victorian Naturalist* 118, 230-31.

Ahern L (2004) Biodiversity Action Planning -Gippsland Lakes Landscape Zone, Gippsland Plain bioregion. DSE/ West Gippsland CMA. [Similarly for LaTrobe, Macalister, Nooramunga, Stradbroke, Tanjil and Yarram Landscape Zones.]

Ahern L (2004) Biodiversity Action Planning—Lower Kiewa Landscape Zone. DSE / North East CMA. [Similarly for Lower Ovens, Mid Kiewa, Mid Ovens and Mid-King Landscape Zones.]

Ahern L and Bennett, AF (1999) Evaluating the proposed Warrandyte- Kinglake Nature Conservation Link. Unpubl. report for Department of Natural Resources and Environment.

Ahern LD and Blyth, JD (1979) Qualitative study of benthic invertebrates—Mitchell River Study. Supplementary Report on Environmental Studies. State Rivers and Water Supply Commission, June 1979.

Ahern LD, Brown PR, Robertson P, Seebeck JH, Brown AM and Begg, RJ (1985) A proposed taxon priority system for Victorian vertebrate fauna. *Arthur Rylah Institute Technical Report Series* 30

Ahern LD, Brown PR, Robertson P and Seebeck JH (1985) Application of a taxon priority system to some Victorian vertebrate fauna. *Arthur Rylah Institute Technical Report Series* 32

Ahern L, Davidson I and Quinlivan G (2007) Biodiversity Action Planning—Lower Mitta Mitta Landscape Zone. North East CMA. [Similarly for Upper Murray, Omeo and Upper Ovens and King Landscape Zones.]

Ahern LD, Frood D and Robertson P (1998) Flora and Fauna Assessment – O'Herns Road, Epping CT979838. Unpubl. report for City of Whittlesea.

Ahern L, Lowe K, Handley K, Berwick S and Robinson D (2001) A plan for conserving native biodiversity in the Northern Inland Slopes Riverina bioregion, Victoria— Strategic overview. Department of Natural Resources and Environment. Victoria; North Central CMA; Goulburn

Broken CMA; North East CMA. Unpublished report.

Ahern L, Lowe K, Handley K, Berwick S and Robinson D (2002) A plan for conserving native biodiversity in the Victorian Riverina bioregion, Victoria—Strategic overview. Department of Natural Resources and Environment, Victoria; North East CMA; Goulburn Broken CMA; and North East CMA. Unpublished report

Ahern L, Lowe K, Handley K, Howell M and Robinson D (2002) A plan for conserving native biodiversity in Victorian bioregions—assets and actions. Goulburn Broken CMA Area 3: 'Shepparton irrigation region'. Department of Natural Resources and Environment, Victoria, and Goulburn Broken Catchment Management Authority. Unpublished report.

Ahern L, Lowe K, Handley K, Park G and Alexander J (2002) A plan for conserving native biodiversity in the Murray Fans bioregion, Victoria—Strategic overview. Department of Natural Resources and Environment, Victoria, Goulburn Broken CMA; North Central CMA; and Mallee CMA. Unpublished report.

Ahern L, Lowe, K, Higgins L, Park G and Diez S (2002) A plan for conserving native biodiversity in Victorian bioregions—assets and actions. North Central CMA Area 1: 'western northern plains'. Department of Natural Resources and Environment, Victoria, and North Central Catchment Management Authority. Unpublished report.

Ahern L, Lowe K, Moorrees A, Park G and Price R (2001a) A strategy for conserving biodiversity in the Goldfields bioregion, Victoria: Strategic Overview. Department of Natural Resources and Environment, Victoria. Unpublished report.

Ahern L, Lowe K, Moorrees A, Park G and Price R (2001b) A strategy for conserving biodiversity in the Goldfields bioregion, Victoria. Volume 2: Zones, assets and actions–All Areas of Goldfields bioregion. Department of Natural Resources and Environment, Victoria. Unpublished report.

Ahern L, Lowe K, Robinson D, Berwick S and Handley K (2002a) A plan for conserving native biodiversity in Victorian bioregions—assets and actions. Goulburn Broken CMA Area 1: 'south-central woodlands'. Department of Natural Resources and Environment, Victoria, and Goulburn Broken Catchment Management Authority. Unpublished report.

Ahern L, Lowe K, Robinson D, Berwick S and Handley K (2002b) A plan for conserving native biodiversity in Victorian bioregions –assets and actions. Goulburn Broken CMA Area 2: 'north-eastern woodlands'. Department of Natural Resources and Environment, Victoria, and Goulburn Broken Catchment Management Authority. Unpublished report.

Ahern LD and Platt SJ (1991). Land for Wildlife – catching Victoria's fancy. Australian Ranger 23, 5-8.

Ahern LD, Tsyrlin E and Myers R (2001) Mount Donna Buang Wingless Stonefly Riekoperla darlingtoni. Flora and Fauna Guarantee Action Statement No. 125. Department of Natural Resources and Environment, Victoria.

Ahern LD and van der Ree R (2002) Squirrel Glider Petaurus norfolcensis. Flora and Fauna Guarantee Action Statement No. 166. Department of Natural Resources and Environment, Victoria.

Ahern LD and Yen AL (1977) A comparison of the invertebrate fauna under Eucalyptus and Pinus forests in the Otway Ranges, Victoria. Proceedings of the Royal Society of Victoria 89, 127-136.

Frood D and Ahern L (2003) Environmental assessments of bridge sites in Closed Catchments within Yarra Ranges National Park – Report 1. Melbourne Water / Parks Victoria (internal report).

Frood D and Ahern L (2004) Environmental assessments of bridge sites in Melbourne Water Closed Catchments – Report 2. Melbourne Water/Parks Victoria (internal report).

NRE (2000) Resource protection guidelines - rabbit control: Mallee. Department of Natural Resources and Environment, East Melbourne.

NRE (2000) Resource protection guidelines - rabbit control: North Central. Department of Natural Resources and Environment, East Melbourne.

NRE (2000) Resource protection guidelines - rabbit control: Wimmera. Department of Natural Resources and Environment, East Melbourne.

Norris KC, Mansergh IM, Ahern LD, Belcher CA, Temby, ID and Walsh NG (1983) Vertebrate fauna of the Gippsland Lakes Catchment, Victoria. Fisheries and Wildlife Division, Victoria Occasional Paper Series 1.

Platt SJ and Ahern LD (1995) Nature conservation on private land in Victoria, Australia—the role of Land for Wildlife. In Nature Conservation 4: The Role of Networks. Ed DA Saunders, J Craig and EM Mattiske (Surrey Beatty & Sons, Chipping Norton)

Platt SJ and Ahern LD (1995) Voluntary nature conservation on private land in Victoria—evaluating the Land for Wildlife programme. In People and Nature Conservation—Perspectives on Private Land Use and Endangered Species Recovery. Eds A Bennett, G Backhouse and T Clark Transactions of the Royal Zoological Society of New South Wales. (Surrey Beatty & Sons, NSW)

Robertson P and Ahern L. (in prep.) A survey and risk assessment of terrestrial vertebrate fauna of the Murray Scroll Belt. Final report of a two-year field study conducted during spring/summer 2004-05 and summer 2005-06, including a compilation of comparable historical data. Wildlife Profiles P/L and Parks Victoria.

Webster R and Ahern L (1992) Superb Parrot Polytelis swainsonii. Flora and Fauna Guarantee Action Statement No. 33. Dept. of Conservation and Natural Resources, Victoria.

Webster R and Ahern L (1992) Management for the Conservation of the Superb Parrot (Polytelis swainsonii) in New South Wales and Victoria. Department of Conservation and Natural Resources, Victoria and National Parks and Wildlife Service, NSW.

Woolley PA and Ahern LD (1983) Observations on the ecology and reproduction of Sminthopsis leucopus (Marsupialia: Dasyuridae). Proceedings of the Royal Society of Victoria 95, 169-180.

Prepared by Leigh's colleagues, including Chris Belcher, Ian Mansergh, Peter Menkhorst, Peter Robertson, Ian Temby, Beverley Van Praagh and Alan Yen.

Thomas Henry Sault OAM

26 July 1922 - 10 September 2008

Tom Sault was born in Eildon to Gladys and Harry Sault. A year later his sister Clara was born. He had only a few years of schooling at the local primary school, where the boys generally went barefoot, before the impact of the Depression resulted in the family moving. Tom's father took on work with a construction crew, building roads and bridges in the Heytesbury Forest near Timboon. There the family lived in a tent, with saplings and wheat or potato bags for beds. Tom's mother cooked for the men on a large wood stove under a tarpaulin. Tom's job was to collect the wood and kindling, as well as the water in four-gallon tins. Once this was done, he often set off barefoot into the bush to explore.

Whilst in the area, Tom and Clara apparently attended the nearby Paaratte School. Occasionally they had an outing to the beach near Peterborough and Port Campbell. Without a doubt, this freedom helped further develop Tom's love of nature, which was to continue throughout his life.

On their return to Eildon about two years later, Tom returned briefly to school. A second house, which had belonged to the neighbours, was moved next to theirs and his father added several rooms onto it. Tom's mother then began a guest house for visitors, particularly fishermen. She provided lodgings with home-cooked meals for 35 shillings a week. Candles, lamps and a fireplace provided light, and water was later pumped up to the house from the river. Harry and his friend took up gold prospecting in the Jerusalem Creek area.

Tom learnt to fish and became a keen fisherman over the years. He also caught rabbits and sold the skins, played the mouth organ, learnt to play the piano, milked the cows and cared for the pigs. On Wednesdays he rode his bike out along rough tracks to see his father, taking him more food. On one occasion, he and his father explored Enoch's Point together, hoping to find another place to search for gold. They pushed everything they needed in a wooden barrow and lived on rabbit stew. However, their trip proved fruitless.

Tom enjoyed having his casement window open, especially at night. He would listen to the river, watch the stars, observe bats coming into his room to catch insects, and enjoy the birds in the early morning.

At age 14, Tom was working on a pine plantation on the Delatite Arm of Eildon Weir, keeping up the supply of seedlings to the men as they planted them on the steep slopes. He was paid two pounds six shillings a week, half the basic wage. Tom trapped rabbits after work and lived mainly on rabbit stew. Later, he stoked the boilers at a timber mill near Snob's Creek. It was pleasant in winter, but terribly hot in summer. Tom worked five and a half days a week and then walked home for the remainder of the weekend, through the bush and along animal tracks. He would return on Sunday night, having enjoyed his mother's cooking and carrying his food for the week ahead.

He worked for the Forest Commission at about age 17, helping to build a fire lookout of

112

Mountain Ash and Alpine Ash on Lake Mountain. He said 'It was the coldest job I ever had. I suffered from chilblains and could hear the dingoes howling at night.'

During World War II, Tom enlisted at age 18 and served in New Guinea, the Solomon Islands and Rabaul. Even there, he had his eyes open for anything related to natural history, which helped ease his wartime experiences.

On his return, Tom knew he did not want to go prospecting. He completed a cabinet-making course in Melbourne through the Post War Scheme for ex-servicemen. He then went on to build many staircases and cupboards for shops and homes in Melbourne.

Later, Tom's outdoor interests resulted in him completing a geology course and this in turn led him to the Field Naturalists' Club of Victoria (FNCV), in 1962. He immediately became actively involved in the Geology Group, followed by the Botany Group and the Mammal Survey Group (now the Fauna Survey Group).

He was a member of Council from 1964 to 1966 and again from 1974 to 1979. Tom was elected Vice President from 1966 to 1969 and in 1970 became President for three years. During the 1960s and 1970s he was heavily involved with the annual Nature Show held at the Melbourne Town Hall, which promoted the FNCV.

Tom participated in Fauna Survey Group activities for many years, organising camps, collating records, transporting equipment, and attending most survey camps and meetings. At the surveys in the Big River valley near Eildon area, where Tom had lived as a boy, his local knowledge was valued, and he was able to take members on an excursion to his father's gold mine to search for bats. Around the campfire he could relate the local history as he knew it to a younger generation. The Fauna Survey group conducted many surveys in coastal woodland and heathland such as on Mornington Peninsula, Western Port, Wilson's Promontory and Nooramunga, in search of rarities such as the New Holland Mouse. It was in these surveys in particular that Tom's knowledge of coastal heathlands came to the fore.

In 1989, the Fauna Survey Group presented Tom with a plaque on a blue gum burl for his dedication to the group over many years. He became an Honorary Life Member of FNCV in 1991.

Tom had retired in 1982, and moved to the Mornington Peninsula. He quickly joined the Friends of Arthur's Seat State Park and went on to be a founding member of several other groups including the Friends of the Hooded Plover, the Friends of the Tootgarook Wetlands and the Seawinds Nursery Volunteers. He was a life member of the Southern Peninsula Tree Preservation Society and the Southern Peninsula Indigenous Flora and Fauna Association.

During the late 1990s, Tom and Clara were taken back to Eildon. The original property had been flooded when Lake Eildon was enlarged, much to their mother's sorrow. She had resided there for fifty years. It gave Tom much pleasure to rekindle old memories and to encounter several people he had known. Their land and the remains of buildings were clearly visible later, as a result of the drought.

The Mornington Peninsula Through the Eye of a Naturalist was published in 2003, recounting Tom's personal observations and knowledge of the plants, fauna and geology of the region over many years.

In 2005 Tom was awarded a Medal of the Order of Australia (OAM) 'for service to the community of the Mornington Peninsula, particularly through the promotion and protection of the natural environment.'

Tom was an avid reader and kept up-to-date with many issues. He was largely self-taught, had a remarkable memory and was extremely interested in how the geology of an area influenced the flora and fauna. Despite ongoing health issues he loved life, had a sparkle in his eye and remained involved, positive and interested in everything the FNCV and the groups on the Peninsula were doing, until he passed away. He was always well-respected, sharing his knowledge of fauna, flora and his own personal experiences generously, patiently and enthusiastically with both young and old. Tom showed great integrity throughout his life and was a wonderful mentor and friend to all who knew him. He will be greatly missed by many people.

Sally Bewsher
2 Ramsay Court
The Patch, Victoria 3792
with assistance from **Ray Gibson**

Linda Margaret Potter

10 December 1919 - February 2009

If one word could describe Margaret Potter it would have to be 'involvement', which makes her death doubly sad in that she died alone during the extreme heat wave that preceded the Black Saturday bushfires. Whatever she undertook she always carried out with dedication, precision and enthusiasm.

Linda Margaret Potter (known to everyone as Margaret, the Linda disguised always under an initial) was born on 10 December 1919. She was educated at Presbyterian Ladies College, returning there to teach, and ultimately becoming senior chemistry mistress. When she retired in 1980 she joined the ex-collegiate group, in which she was very active. Teaching was in her blood, and when her great-nephews Sam and Bennett began school, she was dissatisfied with the way they were being taught to read, and promptly set about making sure they had a good grounding in this skill. She wanted to do the same thing when they started to learn chemistry, but admitted ruefully that teaching methods and the state of knowledge had changed since her day.

On her retirement she also joined the Field Naturalists Club of Victoria (FNCV), being elected in August 1980. She was a Council member from 1983-1986, and in 1985 was elected chairperson of the Botany Group, a position she held until 1990. Margaret was an inspirational leader, and the Botany Group was most active during her tenure.

Meetings were well-attended, and excursions near and far were organised: to places such as Brimbank Park and the Organ Pipes National Park, Kinglake National Park, Lake Mountain, Phillip Island, the Otways and the Grampians, and the Mornington Peninsula, to remove boneseed. This was the period when the campaign to conserve Courtney's Road, Belgrave South, got under way, due to the efforts of members of the Botany Group. The area is now the Baluk Willam Nature Conservation Reserve. When, in 1987, the FNCV received a request to help

Margaret Potter at Maranoa Gardens, 1990

preserve the Knox Horticultural Research land, which was under threat of development, it was Margaret who wrote letters to the relevant minister and local member on behalf of the club.

Margaret was always punctilious about drawing the attention of the Botany Group to relevant matters from Council minutes, and General Meetings, for example, preparing a summary sheet in 1989 on the *Flora and Fauna Guarantee Act*, following a talk on the subject at a General Meeting. She regularly attended General Meetings, providing exhibits and nature notes, often about things she had observed in her own garden or local parks. After an excursion to the Ada

Tree in 1989 she talked to a General meeting about the importance of its preservation, and the meeting approved the sending of a letter to the Department of Conservation, Forests and Lands urging this. In 1990 she coordinated the FNCV display at the Maranoa Gardens Festival.

The late 1980s were troubled times in the club, in particular due to the shortcomings of the Subscription Secretary. In 1991 Margaret volunteered to sort out the confusion that had ensued, and spent eight hours a week for two years dealing with the complexities of the membership records. In recognition of this she was made an Honorary Member. When, in 1993, Council decided to advertise for a Subscription Secretary, Margaret wrote a detailed report on what was involved, pointing out firmly that six hours a week was not enough. For the following three years she acted as the club's Publicity Officer, ensuring that, among other outlets, notice of the club's activities appeared regularly in *The Age*.

Margaret's involvement in the club was more on the organisational side of things, but in 1984 she led a Day Group excursion to Bellbird Dell Reserve bushland, and in 1995 gave a talk to the Botany Group, entitled 'Kosciusko in Summer'. The latter was her last recorded activity, but for some years she was a familiar figure, with her smile and cheerful enthusiasm, at meetings, including the annual Australian Natural History Medallion reception and presentation.

I am indebted to Beryl Fookes for some of the above information.

Sheila Houghton
12 Scenic Court
Gisborne, Victoria 3437

One Hundred Years Ago

The nest of another Scrub-Robin was found by Mr. M'Lennan, who knows the habits of these birds thoroughly. It was a simple cup-shaped structure of sticks and small twiglet into the ground, and contained a solitary egg, the usual clutch of this bird. Our guide has nicknamed this bird "the trapper's companion" on account of its inquisitiveness, sociability, and fearlessness when he has been out quietly setting traps for Dingoes and other vermin. They are fairly tame so long as no noise is made, but directly they hear a sound they disappear' at once, only to reappear when all is still again. Sometimes the nest is built amongst the bark or debris at the base of a mallee bush, and at times is situated quite 18 inches from the ground. When returning to the nest after having been frightened these birds adopt a coquetting action, approaching within a yard or two of it then rapidly darting away into the scrub, only to repeat the same performance immediately afterwards. One has only to possess a little patience and keep quiet and the Scrub-Robin will show exactly where the egg or young one, as the case may be, is located ...

Several nests of the Graceful Honey-eater, *Ptilotis ornata*, were also found, while many Spiny-checked Honey-eaters were seen feeding in the tree-tops. Restless Flycatchers, *Sisura inquieta*, and Red-capped Robins, *Petroeca goodenovii*, were also seen, busily engaged building their nests. Evidence of the presence of Emus, indicated by tufts of feathers, were seen in many places. Unharnessing our horses, we fastened them up, and then went for a long tramp over sand dunes covered with pines' and scrub. The pad led us to a beautifully green oval space of about ten acres, named by Mr. M'Lennan "the Dingoes' recreation reserve." Here we espied a fox, and presently a Dingo. Probably both of these have by this time yielded their scalps to our guide, who claims to have accounted for the deaths of close on 3,000 Dingoes. Several Mallee-Fowls' nests were inspected, and additional notes on the habits of other birds were made. Many Chestnut-rumped Ground-Wrens were seen in this part.

From *The Victorian Naturalist* **XXVI**, p. 70-71, 73, October 1909

Bowerbirds

by Peter Rowland

Publisher: *CSIRO Publishing 2008. 144 pages; paperback; colour photographs.*
ISBN 9780643094208. RRP $39.95.

For anyone who has an interest in birds, or natural history in general, Peter Rowland's latest book *Bowerbirds* is a worthy addition to their library. The first thing that struck me when I was reviewing this book was the incredible amount of work that Peter has put into his research. The text is full of informative and interesting facts, backed with plenty of scientific research.

Considerable thought has gone into the layout of this book, with early chapters covering areas such as classification, morphology, habitat, distribution etc., followed by information relating to each of the Australian bowerbird species. There is also a supplement covering the New Guinea bowerbirds, which makes for fascinating reading. Each chapter relating to the Australian species includes a distribution map, and there are colour photos representing each species, as well as many photos of the various bowers (photos are by various photographers).

Within each chapter the sub-headings are quite handy, making it easier to find specific information about the bird. These sub-headings cover identification, other names, nomenclature, description, vocalisation, feeding and breeding habits, just to name a few. So often, books with this much information are quite laborious to read but this definitely isn't the case with Peter's book.

Apart from being an interesting book about a particular family of birds, in this case the bowerbirds, this is an ideal reference book for anyone trying to identify bowerbird species, differences in juveniles/adults or a particular bower. There is so much more information than in the general bird identification books.

One thing that I was quite impressed by was the fact that the bibliography extends to nine pages. Peter has obviously researched every imaginable scientific paper, journal and book on bowerbirds in his quest to get the facts right.

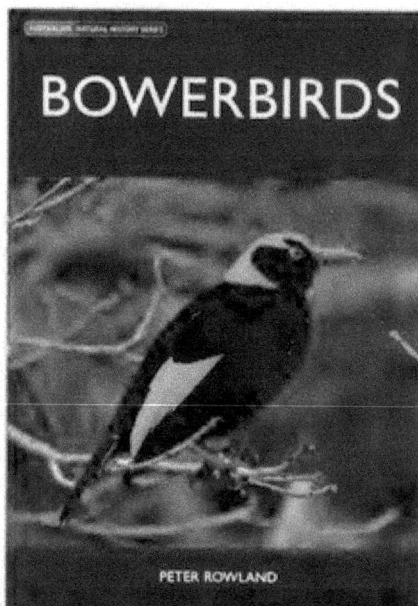

According to the information on the back of this book, Peter Rowland spends much of his time studying and writing about many of Australia's birds. His dedication can be seen throughout this book and I'm sure many birders, naturalists and the general public will appreciate this after reading it. As a natural history writer and nature photographer myself, with a special love for birds, this book will be a welcome addition to my collection.

Michael Snedic
Keperra, Qld 4054
www.michaelsnedic.com

Spirit of the Wedge-tailed Eagle
The Art of Humphrey Price-Jones

Paintings and drawings by Humphrey Price-Jones; text by Penny Olsen

Publisher: *CSIRO Publishing, 2007. 90 pages, hardcover; ISBN 9780643094338, 265 × 355 mm. RRP $50.00*

This large-format, beautifully produced book consists of a collection of paintings and pencil drawings of eagles, accompanied by pertinent quotations from scientific literature. As indicated by the title, the focus is mainly on Australia's largest true eagle, the Wedge-tailed Eagle.

The artist, Welshman Humphrey Price-Jones, has studied and painted Wedge-tailed Eagles ever since he came to Australia forty years ago. He produces life-size portraits of live birds and the original artworks are up to 1800 mm × 1200 mm. Any work smaller than life-size, he maintains, could not convey the majesty and power of these birds. The reproductions in this book, four of which are double-page spreads, are of necessity much smaller, but they still convey the essential characters of these magnificent birds — their strength, noble appearance and bright, keen eyes. The pictures of feathers, skulls and eggs, however, suffer from being shrunk in size because they look too small to belong to such large birds.

Most of the paintings and drawings are grouped into four sections. **Eagles** includes portraits of six of the world's 35 eagles: Verreaux's Eagle (a very close relative of the Wedge-tailed Eagle), Eastern Imperial Eagle, Golden Eagle, Wedge-tailed Eagle, Little Eagle and White-bellied (no longer 'White-breasted' as stated) Sea-Eagle. The latter three species live in Australia. **Wedge-tailed Eagle** includes many portraits of this species perched, shown from a variety of angles. **Eagles in action** depicts these eagles flying, hunting, feeding and being harassed by ravens and Australian Magpies. There is also an aerial view of a bird gliding high over hills and valleys. 'Eaglet to eagle' features pairs of birds, nest-building, and the development of a chick from hatchling to adult.

Raptor expert Penny Olsen has carefully selected the accompanying text from observations recorded in 15 books, mostly by well-known ornithologists such as John Gould, David Fleay,

Alfred J North, Graham Pizzey, Peter Slater, and Penny's own book *Wedge-tailed Eagle*. She has also added some explanatory passages. The informative text covers many aspects of the eagle's physiology and life cycle, including eyesight, feathers, flight, hunting, prey and reproduction.

By the end of the book, you feel that you are getting to know these awesome birds and the challenges they face in order to survive. This book will appeal to raptor enthusiasts and anyone interested in wildlife art, but everyone should at least have a look at it.

Virgil Hubregtse
6 Saniky Street
Notting Hill, Victoria 3168

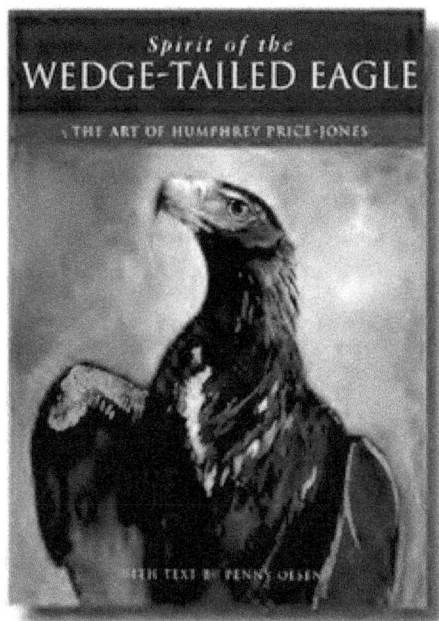

Mound-builders

Darryl Jones and Ann Göth

Publisher: *CSIRO Publishing, 2008. 119 pages, paperback, colour photographs.*
ISBN 9780643093454 RRP $39.95

Another interesting book in the Australian Natural History Series, the current volume is about birds that build mounds to incubate their eggs. This new book covers much of the reproductive and conservation biology of the three species of megapode found in Australia (Australian Brush-turkey, Malleefowl and Orange-footed Scrubfowl). Chapters are titled: Familiar yet distinct; Taxonomy, Distribution and habitat; Appearance and ecology; The mound; Abandoned eggs; Growing up without parental care; Social and reproductive behaviour, Conservation and management of Australian mound-builders. The bulk of the text covers the unique breeding mechanisms of these birds.

The authors do not clearly state the intent of their book in the preface or opening chapters, although the back cover indicates the book is '... an excellent introduction to one of the most unusual bird families'. It isn't until page 5 that the authors suggest they hope to describe progress in understanding megapode species since Harry Frith's seminal work on the Malleefowl in New South Wales (published 1962). Readers are most likely to be familiar with the Brush Turkey and Malleefowl, and this is where the book will be of most interest to the general naturalist or birdo.

Jones and Göth are eminently qualified as experts in megapode biology. They come from a university research background and both have published a number of papers and books in birding literature, including several on mound-builders. In producing this book, the authors surveyed much scientific literature and undertook field work in their own megapode studies. The chapter topics cover the group's biology very well (with the exception of the Scrubfowl of which there is still limited knowledge).

There are many interesting snippets. This reader was unaware of the apparent relationship between the exotic Prickly Pear distribution and populations of the Brush-turkey during the spread of this weed (p. 14), the glow of the Brush-turkey wattle (p. 19) and the fact these birds can eat some poisonous plants (*Alocasia* spp. p. 28). Possibly better known is that the Malleefowl has been used to predict weather changes (rain, p. 47) and that mound temperature affects the ratio of male and female Brush-turkey fledglings (p. 64). Göth's studies have produced many interesting findings on Malleefowl and Brush-turkey biology (p. 78).

I had few quibbles with the book. As with some of the previous volumes in this series, proofing of the text was variable. Australian Bustard has an incorrect scientific name (p. 13, should be *Ardeotis australis*). The distribution map for the three megapode species (p. 13) appears accurate for the Brush-turkey and Scrub-fowl but mystifying for the Malleefowl. Many of the South Australian populations shown for the latter do not appear in any of the bird atlases currently published. Nor do all the SA populations appear on maps in the current Malleefowl Recovery Plan (see references). A glossary of terms would have been useful as not all readers would be familiar with of the terms used (e.g. precocial vs. altricial young, p. viii).

The most interesting parts of the book are those covering the features that make mound-builders unique, i.e. the breeding and conservation biology and the fascinating mound-building behaviour. The book succeeds in conveying these aspects of Malleefowl and Brush-turkey biology. The lack of knowledge on the biology of the Scrubfowl is apparent and deserves more attention by researchers.

Any reader wishing to learn more about Australian mound-builders in an accessible format will find this book of interest. It is short enough to read all or part of the book in just a few sittings. It increased my knowledge and understanding of this family and certainly highlights the unique biology of these birds. Anyone with an interest in bird behaviour will find the book interesting.

Martin O'Brien
Department of Sustainability and Environment,
PO Box 500, East Melbourne 3002

References
Frith, H (1962) *The Mallee-Fowl: The Bird that Builds an Incubator*. (Angus and Robertson: Sydney)
Malleefowl Recovery Plan (http://www.environment.gov.au/biodiversity/threatened/publications/recovery/malleefowl/index.html)

One Hundred and One Years Ago

PROTECTION OF NATIVE BIRDS — Among other letters on this subject which have appeared in the Argus lately was the following forcible one from Mr. G. E. Shepherd, of Somerville, an enthusiastic ornithologist. He says :— "The thanks of all nature lovers, particularly ornithologists, are due to you for your very able and opportune article regarding our indigenous birds. As a resident of Mornington Peninsula for upwards of 40 years, I say most emphatically that even now the result of the indiscriminate destruction of birds is beginning to be felt. Lagoons and swamps that were considered, to be permanent 40 years ago are dry depressions, as a result of the wading birds that kept the yabbies in check being either driven away or slaughtered. Only two seasons ago, whilst making bird observations in and around a lagoon, I noticed a stately Pacific Heron feeding in the shallow water. My successive visits seemed to inspire confidence in this noble creature, but, alas, less than a week elapsed ere I found him dead on the margin of the swamp, shot merely for amusement. The White-fronted Heron consumes large quantities of grasshoppers and crickets. I have seen the birds working in hundreds in a potato field, coming in the early morning, and remaining all day, retiring to thick timber to roost in the evening. White Herons are now very scarce, the Bittern and Nankeen Night-Heron are seldom seen, and, unfortunately, when seen are very often shot, like the heron previously mentioned. Hawks are beginning to become very scarce here, a result largely due to people's ignorance. Even the beautiful and harmless little Kestrel is shot " on sight," simply because it is a hawk, without a single thought being given to the fact that it has its own field to labour in, and its own destiny to fulfil. To the State schools and teachers we must, I think, look for the remedy. Let children be taught that it is wicked to destroy birds without good reasons ; also let them be taught to see for themselves that bird-life is part of the great scheme of Nature. Finally, let us have laws enacted and administered that will be a protection to useful birds of all classes."

From *The Victorian Naturalist* XXV, p. 88, October 8, 1908

Albatross: their world, their ways

Tui De Roy, Mark Jones and Julian Fitter

Publisher: *CSIRO Publishing, 2008. Large format, hardback 232 pages.*
ISBN 9780643095557 RRP: $79.95

This is a beautifully presented, larger format, hardcover book, something that is both a pleasure to browse and read. The bulk of the photographs are by Tui de Roy and these are superb; albatross with sunsets and albatross in panoramas that capture the spirit of the wild breeding grounds are some of my favourites.

Albatross is divided into three sections. The first, by de Roy, details each species or species group as a photographic essay. The combination of images and text provides insights into de Roy's experiences whilst visiting remote island breeding grounds. Although there is a bias towards New Zealand sites, this is not inappropriate given 11 forms of albatross breed there. Parochial Australians may, however, be disappointed with the treatment of the only form to breed in close proximity to this continent: a single paragraph with no accompanying photographs indicates de Roy did not connect with the Shy Albatross *Thalasache cauta* (*cauta*) at one of Albatross Island, The Mewstone or Bedra Branca.

The second section deals with science and conservation. Here, many of the world's leading albatross researchers and conservators have contributed short essays that capture lifetimes of experience. While some have been instrumental in directly reducing the slaughter of albatross through unacceptable fishing practices, others have contributed to knowledge of at sea ecology, breeding systems, flight energetics and so on. These are interesting essays; the more so for their diversity and the origins of the contributors. Take for example Conrad J. Glass; a police inspector, conservation officer and direct descendant of founders two centuries prior on Tristan da Cunha, the world's remotest inhabited island. His contribution is an elegant summary of the impact of man since the discovery of this volcanic outpost in 1506, and the generational changes that have ultimately seen a once staple food source now fully protected. Who knew that the Sooty Albatross was a legal component of the menu until 1986?

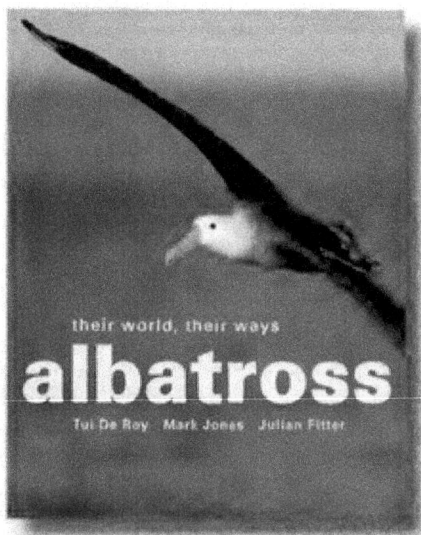

Australian contributions also feature here. Rosemary Gales, a Tasmanian based researcher, provides a concise summary of the global conservation status of albatross. It is sobering to cast an eye over population trends for the 22 recognised species of this text. Based on repeated counts at the breeding grounds, 12 remain in decline despite a reduction in the slaughter by fishing industries. The fact that half of the world's albatross populations are going backwards must surely be alarming! The populations for just five species are thought to be stable, though four of these remain threatened, and for a further four species, there remain gaps in knowledge that prevent any accurate assessment. It is therefore ironic that the Short-tailed Albatross is the only species whose population is now increasing. Hunted to 'certain extinction' for its plumes, a handful of pairs were re-discovered on an active volcano off Japan in the 1950s. With intensive management, the population has crept upwards to number some

400 pairs today. Such a remarkable resurrection provides hope for other albatross forms whose more recent declines have been precipitous.

The final section is devoted to species accounts. This is the ready reference section of the book, a place for readers to dip into for an identification feature, to brush up on the threats faced by a particular species or to check a distribution map.

Most of my criticisms relate to what's not included rather than problems with what is. For example, while the images that are included are first-rate, they are almost entirely from breeding grounds; albatross on land; feeding chicks; displaying or flying over nesting sites. As a consequence, the images fail to capture the true essence of the albatross – these are creatures of the open ocean 'that expend 95% of their existence at sea' (p. 21). A greater emphasis on photographs of birds at sea, hanging effortlessly in the wind beside a pitching ship, soaring over huge swells or skimming the millpond of a strangely calm ocean would have addressed this. Similarly, because of the bias to the breeding grounds, an entire age-cohort, the far-ranging juvenile and immature birds – of all species – is missing. Beware the birder that spies an albatross from a headland or partakes in what is one of the great wildlife experiences, an organised pelagic excursion. A goodly proportion of the albatross encountered, and those that often pose the greatest challenge to identification, are simply not represented pictorially (and receive scant and largely inadequate treatment in the text). Even with a ready reference section, this is not a field guide nor a book that one takes to sea. Rather, it seems most appropriate that it be read by a warm fire, when 'miserable' weather and howling winds keep one indoors.

Books of this nature are rarely going to do everything, and so, despite these gaps, this remains both a beautiful and informative work. As a 'pelagic tragic' it readily finds a place on my bookshelf and I would thoroughly recommend it to others both as a beautiful gift and a worthy addition to one's library.

Albatrosses

Terence Lindsey

Publisher: *CSIRO Publishing, 2008. 152 pages, paperback, colour photographs.*
ISBN 9780643094215. RRP: $39.95

Albatrosses is the latest addition to the Australian Natural History Series to be produced by CSIRO Publishing. Like others in the series, it aims to present a comprehensive and up-to-date account of a faunal group in 'a style suitable for upper secondary or undergraduate level readers, as well as (field) naturalists.' At first glance, this book didn't grab me. It has a nice enough cover, and contains 18 colour photographs along with a smaller number of black and white images, sketches and tables, but with page after page of text, and few sub-headings, it isn't a book that one can readily dip into. This will surely make it harder to sell to the general reader and especially the target audience. Don't be put off though, because one must digest this book chapter by chapter to appreciate it. Indeed, once started, I found that it was both interesting and engaging, and not being particularly long, with 112 pages of text and images, I read it in an afternoon.

Albatross are truly remarkable birds and the eight chapters of this book ensure these ocean wanderers are more readily accessible. The somewhat loosely titled first chapter, 'Myth and Legend' is more about setting the scene and enticing the reader with some of the more remarkable aspects of albatross ecology than about the tales and beliefs of seafarers of yesteryear. Regardless, it's a worthy start. This is followed by a chapter outlining the four albatross groups, namely, the great albatross, northern pacific albatross, southern mollymawks and sooties. An introduction for beginners, it contains a blend of descriptive features and ecological characteristics that separate one group from another. This is not a field guide section and never pretends to be; rather, it serves to introduce the reader to the key groups before delving further into albatross ecology. Chapters on the 'Southern Ocean', 'Food and foraging', and 'Flight' combine to provide insights into both

ALBATROSSES

TERENCE LINDSEY

how and why these birds routinely move over such vast distances. Several further chapters deal with 'Courtship' and the remarkably long 'Nesting cycle'. Approaching the final chapter, the reader has an appreciation of the complexities of albatross life history; notably, the degree to which specialisation leaves these birds so exposed to environmental change. With the scene set, this last chapter deals with 'Human impacts'; well written, it can be summed up in single word – depressing.

Lindsey has done a very good job of condensing what is known of albatross (based on thousands of pages of research writings) into a readable format. One aspect I particularly like is his readiness to identify the grey areas and remaining gaps in knowledge, as for albatross there remain many. It was here as a nitpicking reviewer that I was hamstrung at most turns. For example, the text states that 'Wanderer's are the biggest albatross. Nothing else comes close' and I immediately thought 'ahaa, I've seen evidence that suggests Southern Royals take that prize', but reading on, Lindsey writes 'except another albatross. The fact is the two great albatrosses, the wanderer and the royal, are so closely matched in size...'

All in all, this is a book that should be well received by field naturalists. Lindsey's style ensures that the results of quite complex science, based on decades of study, has been condensed into a very readable text.

Dr Rohan Clarke
School of Biological Sciences
Monash University, Clayton, Vic 3800

One Hundred and One Years Ago

A NIGHT WITH THE BIRDS OF LAWRENCE ROCKS
BY A.H.E. MATTINGLEY, C.M.Z.S.

The Dove-like Prion is vernacularly known at Portland as the "Snow-bird." There were very few of their rat-like burrows in this small area of soil, which was riddled in every direction with Penguin and Mutton-bird holes, and as the Dove-like Prion is a fragile bird, and unable to fight either the Mutton-bird or Penguin for its choice of a nesting site, it has perforce to utilize that portion of the rookery unoccupied by these last-named birds, which is the outer edge of the soil where it meets the rock. As the soil, especially at these parts, is loose and friable, the hurricanes that at times come raging over this exposed islet tear away the edges of the rookery and destroy these unfortunate birds. Evidences of the destructive work of wind and water were plainly visible. All along the extreme edge of the rookery were burrows f the Dove-like Prions, from which the covering of soil had been swept away by the wind, whilst in the nesting cavity at the extremity many broken and a few unbroken eggs were found, one egg comprising a clutch, whilst some of the adult birds had been blocked in their burrows and had been smothered. Most of the burrows of these birds had a turn in them, instead of being excavated straight into the soil. This turn was no doubt made by the birds mainly to prevent the complete choking up of their burrows by particles of wind-driven soil, but in some cases the turn in the tunnelling was due to a hard piece of rock intruding and barring the way, rendering it necessary to turn off in another direction.

From *The Victorian Naturalist* XXV, p. 15, May 7, 1908

Boom & Bust:
Bird stories for a dry country

edited by L Robin, R Heinsohn
and L Joseph

Publisher: *CSIRO Publishing. 2009. 312 pages,
hardcover. ISBN 9780643090668 RRP $39.95*

For anyone who has examined ornithological
books, it quickly becomes apparent that a lot
of space is dedicated to the great bird migra-
tions which can often be tracked with precision
through space and time. It is more difficult to
find detailed treatments of bird movements
which are nomadic and irruptive. This book
serves well to fill many of these gaps by examining the less predictable and 'messier' end of
the bird movement spectrum, which character-
ises a significant proportion of the Australian
avifauna.

Included are chapters that examine the life
strategies and responses of relatively common
and widespread birds, such as Zebra Finches,
Grey Teal, Australian Pelicans, woodswallows
and choughs, to unpredictable conditions in
the arid and semi-arid zones of our continent.
We often take these birds for granted, but these
chapters ensure we will not look at them the
same way again. Other chapters cover more
enigmatic species such as the Night Parrot,
examine the palaeoecology and past climate
of Australia by exploring fossil bird evidence
and cover indigenous perspectives of faunal
responses that indicate fluctuating seasons and
climate. Surprisingly, there was only passing
reference to the conservation challenges facing
birds with nomadic and irruptive life histories
and the potential impacts of recent human-in-
duced climate change.

By and large, the chapters are very well written and self contained, meaning that they can
be read in isolation. A minority of chapters
could have benefited from more thorough editing to eliminate typos, awkwardly-written passages and some misinterpretations of scientific
literature. The entire book is well indexed.

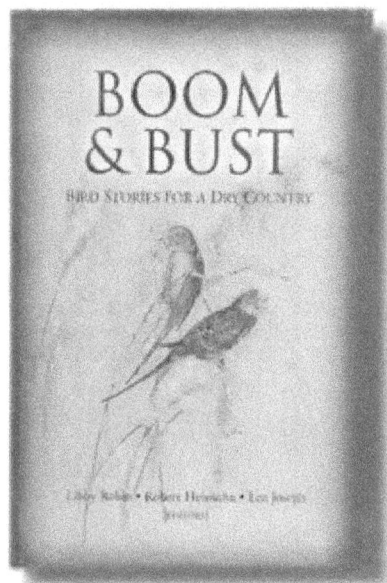

This book makes use of endnotes at the con-
clusion of each chapter which list relevant
references but also act to provide additional
information and anecdotes of varying levels of
relevance to the text. This system unnecessarily
lengthens the book (a separate Select Bibliography is also provided). For instance, one chapter has 24 pages of main text and 13 pages of
endnotes in reduced font and spacing. A more
traditional system of referencing, as used in scientific literature, which was linked to a single
reference section at the rear would have been
preferable.

The presentation of the book is very simple
and elegant. Despite being a hardcover (and en-
cumbered with the endnotes mentioned above)
it remains a very compact and portable volume,
making it a perfect travelling companion to remote regions where conditions are unpredictable and the cycle of boom and bust prevails. I
am sure that anyone who reads this book will
learn something new before putting it down.

Mark Antos
Parks Victoria
535 Bourke St, Melbourne

The Victorian Naturalist
is published six times per year by the

Field Naturalists Club of Victoria Inc

Registered Office: FNCV, 1 Gardenia Street, Blackburn, Victoria 3130, Australia.
Postal Address: FNCV, Locked Bag 3, Blackburn, Victoria 3130, Australia.
Phone/Fax (03) 9877 9860; International Phone/Fax 61 3 9877 9860.
email: admin@fncv.org.au
www.fncv.org.au

Address correspondence to:
The Editors, *The Victorian Naturalist*, Locked Bag 3, Blackburn, Victoria, Australia 3130.
Phone: (03) 9877 9860. Email: vicnat@fncv.org.au

Yearly Subscription Rates – The Field Naturalists Club of Victoria Inc

Membership category		*Institutional*	
Single	$65	Libraries and Institutions	
Concessional (pensioner/Senior)	$50	- within Australia	$120
Family (at same address)	$85	- overseas	AUD130
Junior	$18		
Additional junior (same family)	$6	Schools/Clubs	$65
Student	$25		

(These rates apply were set on 1 October 2008)

All subscription enquiries should be sent to
FNCV, Locked Bag 3, Blackburn, Victoria, Australia 3130.
Phone/Fax 61 3 9877 9860. Email admin@fncv.org.au

www.ingramcontent.com/pod-product-compliance
Lightning Source LLC
Chambersburg PA
CBHW022105210326
41519CB00056B/1447